W9-BRP-523

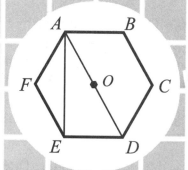

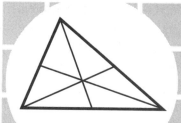

MATH OLYMPIAD
CONTEST PROBLEMS
VOLUME 2

RICHARD KALMAN, EDITOR

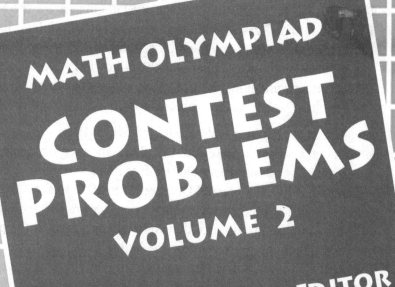

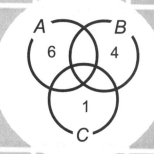

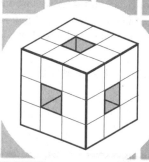

About the Math Olympiads

The Mathematical Olympiads for Elementary and Middle Schools (MOEMS) was created by Dr. George Lenchner, then the Director of Mathematics for the Valley Stream High School District and Consultant to the three associated Elementary School Districts on Long Island, NY. He created four pilot Olympiads for the elementary schools in 1978-79 before formally organizing MOEMS for all Long Island elementary schools in the fall of 1979. Serving as its first Exectutive Director until his phased-in retirement in 1996, he oversaw the growth of MOEMS into an international competition. Dr. Lenchner wrote all the problems for the first 16 years.

His successor, Richard Kalman, joined MOEMS in 1994 to become the second Executive Director after a one year training period. Like Dr. Lenchner, he wrote all the problems, but after three years, he created large committees chaired by Grand Duffrin to create and polish them. Under his leadership, MOEMS created Division M in 1998 to serve students in grades 7-8, allowing its veteran mathletes to continue competing after grade 6, and introduced the naming of strategies for all problems and providing extensions for about half of the problems. Other changes created during his tenure include two Websites, one highly informative for the general public and the other a secure two-way communication system for MOEMS coaches to replace the mails; a strong presence at mathematics teachers' conference; professional development workshops in problem solving, regional mathematics tournaments separate from the Olympiads, and an increase in the number of foreign affiliates.

Publisher:

Mathematical Olympiads for Elementary and Middle Schools, Inc., Bellmore NY, USA
Website: http://www.moems.org   Phone: 516-781-2400

Editors:

Grant Duffrin, contest editor
Richard Kalman, book editor

Layout, Graphics, and Cover Design:

Valerie Cevallos

Printer:

Tobay Printing Company, Inc. Copaigue, NY

Library of Congress Catalog Number: 2007908880

ISBN: 978-1-882144-11-2

Dr. George Lenchner
(1917 - 2006)

This book is dedicated to the memory of Dr. George Lenchner, who created the Math Olympiads out of his pure love for students and learning. His entire career was an inspiration for thousands of teachers and students.

TABLE OF CONTENTS

Preface

Learning to compute efficiently is the lesser part of learning mathematics, although in the earlier grades the focus is primarily on arithmetic. Learning to reason logically is the major skill students must develop. Learning to solve authentic problems is the most important tool students can use toward this end.

This book is a collection of 425 challenging and interesting problems authored for the *Mathematical Olympiads for Elementary and Middle Schools* (*MOEMS*) over a ten-year period from 1995 to 2005. Over 150,000 students participate every year, representing about 6500 teams worldwide. About two-thirds are within the United States.

The original form of the problems in this book were created by J. Bryan Sullivan (seven Division E sets), Gene Vetter (six Division M sets), and Richard Kalman (four sets). Additional problems were contributed by Mary Altieri, Curt Boddie, Dr. George Lenchner, and Jerry Resnick. The Problem Writing Committee, which polished the questions and wrote the solutions, was chaired by Grant Duffrin, who also performed as contest editor. Special appreciation is due all of the above for their masterful contributions to the MOEMS contests.

Committee members who contributed significantly to the polishing and previewing of the contest problems over the years are: Mary Altieri, Carol Babcock, Dr. Elliott Bird, Curt Boddie, Sandy Cohen, James Connelly, Michael Curry, Barbara Duffrin, Grant Duffrin, Richard Kalman, Gilbert Kessler, John Lufrano, Mary Ann Mansfield, Dennis Mulhearn, Jason Mutford, Eric O'Brien, Oana Pascu, Jerald Resnick, George Reuter, Marj Rubin, Robert Saenger, Arthur Samel, Marilyn Tahl, Jean Wahlgren, and Lawrence Zimmerman.

To them we owe great thanks for their skill and judgment in making recommendations with respect to appropriateness, problem placement, level of difficulty, clarity, and ambiguity.

I am also indebted to the following people for their valuable (or is it invaluable?) review of all or part of the manuscript for this book: Dr. Elliott Bird, Kerrilyn Cierzo, Grant Duffrin, Dr. Carol Findell, Wendy Hersh, Dorothy Hess, Hermine Kalman, John Lufrano, and Dot Steinert.

We all enjoyed creating the problems and solutions and increasing our insights into mathematical principles significantly in the process. I hope you experience the same in testing yourself against them.

Richard Kalman
December, 2007

CONTEST PROBLEM TYPES

Many but not all contest problems can be categorized. This is useful if you choose to work with several related problems even if they involve different concepts.

KEY: the lists below organize the problems by type and are coded by page number and problem placement on that page. For example, "Long division: E (36BD, 43C); M (84A)" refers to three problems in Division E and one in Division M, all involving long division. The three Division E problems are questions B and D on page 36 and problem C on page 43, and the Division M problem is question A on page 84.

A
Addition patterns — see *Number patterns*
Age problems: E (46B); M (109E)
Algebraic thinking: E (24D, 25E, 26C, 27E, 28C, 30B, 32D, 34E, 35D, 37C, 48AB, 53E, 54C, 57AE, 59AC, 61A, 62A, 64CD, 65C, 70B, 71B, 73B); M (79B, 81CE, 84A, 85D, 86B, 88B, 89C, 90D, 91B, 92C, 95C, 98D, 99DE, 103A, 104AD, 105AD, 106A, 108B)
Area: E (27B, 30D, 33D, 34E, 41D, 43C, 48D, 50D, 55E, 58D, 59D, 61D, 62D, 63D, 66E, 71D, 72D); M (76E, 77E, 80E, 81B, 82C, 85B, 87E, 90C, 92E, 94C, 95D, 98E, 102E, 103E, 104E)
—— **and perimeter:** E (31D, 39B, 44D, 45D, 46D, 49E, 52B, 57D, 60D, 64D); M (100D, 101E, 107D)
Arithmetic operations and properties: E (24AD, 25D, 29A, 30AC, 31A, 33B, 34A, 35A, 42A, 49A, 54A, 56C, 59A, 61D, 66A, 69A); M (78E, 81C, 86D, 86A, 89A, 91A, 92D, 93A, 97C, 99B, 102A, 104B, 106A, 107A, 109AD, 110D)
Arithmetic sequences and series — see *Number patterns*
Averages (arithmetic means): E (29E, 36A, 46E, 52C); M (76B, 92A, 96D, 97E, 102B, 103B, 107B, 109C)
—— **Weighted:** E (67E); M (81D)

B
Binary numbers: E (37D)
Business problems: E (39C, 47B, 65B); M (82D, 89D, 91C)

C
Calendar problems: E (65D, 68A); M (84C, 86A, 100E, 104C)
Certainty problems: E (57C); M (87A)
Circles: E (24C, 56D); M (79E, 80AE, 84E, 90C, 94C, 98E, 106D)
Clock problems: E (26A, 28D, 31C, 42C, 46C, 52A, 71E); M (83D, 91E, 110E)
Coin problems: E (28C, 45B, 51E); M (98B)
Combinations: E (24C, 30E, 56B, 63C, 64B); M (80A, 84E, 106B)
Congruent figures: E (26B, 29B, 30D, 39D, 40D, 44D, 45D, 47C, 49E, 51B, 53D, 62D, 65E, 68D, 69BD); M (76E, 77A, 81B, 82C, 90C, 92E, 93D, 94C, 97E, 98E)
Consecutive numbers: E (36D, 38B, 60A); M (83B, 89C, 98AC, 100C, 110D)
—— **Consecutive odd or even numbers:** E (31E, 39E, 107B)
Cryptarithms: E (25C, 33C, 38D, 43A, 47D, 48C, 50B, 55C, 57B, 63E, 72C); M (82B, 84D, 85E, 88C, 95A, 97A, 104A, 106C)

Cubes and rectangular solids: E (25B, 33A, 47C, 50D, 58D, 62D, 63D, 65E, 68D, 69B, 70E); M (76A, 108A)

—— **Painted cube problems:** E (36E, 40D, 53D)

Cycling numbers: E (26A, 32C, 34C, 36C, 42E, 47A); M (77B, 90B, 100E, 102D, 103C)

D

Decimals — see *Fractions*

Digits: E (32C, 36C, 39A, 41A, 42AE, 45C, 46B, 47A, 49C, 50A, 51A, 53B, 56A, 59B, 60E, 62B, 64E, 71C); M (77D, 78D, 81A, 82A, 83B, 85A, 89A, 90B, 92B, 93B, 94A, 95E, 96E, 101A)

—— also see *Cryptarithms and Divisibility*

Distance problems — see *Motion problems*

Divisibility: E (31B, 33E, 36BD, 37A, 43D, 58E, 63E, 66B, 68B, 70A, 71C); M (78AB, 80C, 84D, 85E, 87C, 90E, 96A, 99B, 107C, 108C, 110AD)

E

Even vs. odd numbers — see *Parity*

Exponents: E (32C); M (76D, 91A, 95BE)

F

Factorials: M (82E)

Factors: E (27C, 28A, 32A, 38D, 43E, 72B); M (76D, 79D, 82AE, 84B, 85E, 102C, 103B, 110A)

—— **common factors:** E (31A, 85B, 86D)

Fibonacci numbers: M (103C)

Fractions, decimals, percents: E (24D, 26CE, 27A, 28E, 30A, 33E, 41B, 52E, 66D, 73E); M (78E, 83C, 86CD, 87BD, 89D, 90A, 91BC, 92D, 93A, 94BE, 100B, 101C, 102A, 104D, 105DE, 106E, 108BE, 109C, 110C)

L

Logic: E (34B, 38E, 44B, 45C, 48E, 49C, 50AC, 51A, 52D, 53C, 54E, 55A, 56AE, 58B, 59B, 60B, 61C, 69C, 70C, 71A, 73D); M (78C, 79A, 80B, 81A, 94A, 97D)

M

Motion problems: E (27D, 35C, 45E, 69E); M (88D, 93E, 98D, 99D, 106E)

Multiples: E (55D, 63B, 72A); M (76B, 92A, 105B)

—— **Common multiples:** E (29C, 31B, 34D, 39C, 43D, 44C, 49C, 51D, 54D, 59E, 70A, 71C); M (77B, 99B, 102D)

N

Number patterns: E (24AB, 31E, 34C, 37B, 38C, 39D, 42D, 51BC, 53A, 54B, 61B, 62CD, 64E, 65D, 69D, 70B, 72E); M (90D, 93D, 97C, 101A, 103C, 104C)

O

Odd vs. even numbers — see *Parity*

Organizing data: E (24C, 29A, 30E, 31E, 37E, 41AC, 43B, 49ABD, 51C, 54A, 56BC, 60E, 61E, 63C, 64B, 68C); M (76C, 77D, 78BD, 83B, 85A, 86E, 92B, 93B, 94A, 96CE, 101D, 105E, 107E)

P

Palindromes: M (92B, 96B)

Parity (*odd vs. even numbers*): E (36D, 40A, 49C, 62B, 63B, 67B, 70A); M (77C, 78AB, 82D, 84B, 90E, 94A, 97B, 98C, 103C)

Paths — see *Taxicab geometry*:

Perimeter: E (32B, 35B, 47C, 67D, 73C); M (77A, 79E, 99C, 106D, 108D, 109B)

—— also see *Area and perimeter*

Prime numbers: E (31A, 67B); M (77C, 79D, 82E, 97B, 103D, 109D, 110AD)

Probability: E (72B, 73E); M (93B, 108C)

Process of Elimination: E (33C, 40AD, 52D, 66B); M (82D, 104B)

R

Ratios and proportions: E (27D, 29C, 32B, 37C, 40C, 46A, 54C, 55B, 58C); M (80D, 81E, 88D, 93C, 95C, 99AE, 103E, 110C)

Rectangles and squares: E (27B, 28B, 30D, 31D, 32B, 33D, 35B, 39B, 41D, 43C, 44D, 45D, 46D, 48D, 49E, 52B, 55E, 57D, 58D, 59D, 60D, 61D, 63D, 64D, 66E, 67D, 71D, 72D, 73C); M (76E, 77AE, 79E, 80E, 81B, 82C, 85B, 87E, 89E, 92E, 94C, 99C, 100D, 101E, 102E, 104E, 106D, 107D, 108D, 109B)

Rectangular solids — see *Cubes and rectangular solids*

Remainders: E (25D, 26D, 34C, 37A, 42E, 44C, 52A, 59E, 62E, 63D, 68E, 72A); M (78A, 86A, 89B, 96A, 104E, 107C)

—— also see *Calendars*

S

Sequences — see *Number patterns*

Shortest paths — see *Taxicab geometry*

Signed numbers: E (47E); M (88A, 94B, 95B, 99A, 100C)

Squares — see *Rectangles and squares*

Square numbers: E (31E, 42D, 73A); M (79D, 93D, 105B, 107E, 109D)

T

Target problems: E (35E, 40A, 61E); M (76C, 104B)

Taxicab geometry: E (37E, 70E); M (83E, 88E)

Terminal zeroes: E (28A, 43E); M (82E)

Tests of divisibility — see *Divisibility*

Tower problems: E (26B, 29B, 39D, 40D, 51B)

Triangle inequality: M (105C)

Triangles: E (27B, 30D, 59D, 66E, 69D, 71D, 72D); M (77E, 80A, 81B, 83A, 86E, 93D, 95D, 96C, 97E, 101A, 103E, 105C)

Triangular numbers: E (24C, 29B, 30E, 34C; 51B, 63C, 64B, 67A); M (79A, 85A, 90D, 101A)

V

Venn diagrams: E (29D, 66C); M (79C, 86C, 94D, 108C)

Volume: E (25B, 39D, 47C, 58D, 65E, 68D)

W

Well-known problems:

    Asterisk Array problem, the: E (31E)
    Clock-Angle problem, the: E (91E)
    Ducks problem, the: E (39E)
    Interesting Date problem, the: M (82A)
    Math Olympiad problem, the: E (77B)
    Number Recycling Machine problem, the: E (36C); M (90B)
    Quiz Game problem, the: M (91D)
    Triangle Inequality problem, the: M (105C)
    Wandering Pet problem, the: M (80E)

Working backwards: E (28E, 30A, 37B, 41B, 45A, 47E, 58A, 65B); M (79B, 84A, 103A, 105D, 110B)

INTRODUCTION

For the Reader

This book was written for both the participants in the Mathematical Olympiads for Elementary and Middle Schools and their advisors. It is suitable for mathletes who wish to prepare well for the contests, students who wish to develop higher-order thinking, and teachers who wish to develop more capable students. All problems were designed to help students develop the ability to think mathematically, rather than to teach more advanced or unusual topics. While a few problems can be solved using algebra, nearly all problems can be solved by other, more elementary, methods. In other words, the fun is in devising non-technical ways to solve each problem.

The 425 Math Olympiad contest problems contained in this book are organized into 17 sets of five contests each, every set representing one year's competition. Ten of the sets were created for Division E, which originated in 1979 for students in grades 4-6, and the other seven for Division M, which was added in 1998 for students in grades 7-8. These problems exhibit varying degrees of difficulty and were written for contests between 1995 and 2005, inclusive.

The introduction is arranged into three parts. Sections 1 through 5, written for all readers, contain discussions of problem solving in general. Sections 6 through 8 offer many suggestions for getting the most out of this book. Sections 9 through 14, designed for the advisor, called the **Person-In-Charge-of-the-O**lympiads (PICO), include recommendations related to the various aspects of organizing a Math Olympiad program.

1.   How to Use This Book

Establishing a Study Schedule
A little learning every day is more effective than large chunks of learning once a month for two reasons. The mind needs time to absorb each new thought, and constant practice allows frequent review of previously learned concepts and skills. Together, these foster retention. Try to spend 10 or 15 minutes daily doing one or two problems. This approach should help you minimize the time needed to develop the ability to think mathematically.

You might want to track growth over time by recording success rates for these problem sets. Since you are probably changing the way you think about mathematics, your growth needs time to become apparent. Before long, you are likely to find solving problems intensely and increasingly rewarding.

Choosing Problems

What criteria can help you choose problems from this book? You might pick those that appeal to you, or those of a specific type, or you might go through all contests in order. You might want to select the easiest problems or the most difficult. Section 6 contains suggestions to help you select problems according to need and preference. Section 7 offers ways to get started on those problems that stump you. Section 8 describes the most commonly used strategies. Ultimately, you should test yourself against each and every problem. If you are stuck, see the hint for that problem on pages 112-128.

Using the Solutions in This Book

Whether you solve a problem quickly or you are baffled, it is worth studying the solutions in this book, because often they offer unexpected insights that can help you understand the problem more fully. After you have spent time — actually, invested time — trying to solve each problem any way you can, reviewing our solutions is very effective. Think of each problem as a small doorway that opens into a large room of mathematical thought. You want to be in that room. There you will find a wealth of concepts bundled together into one or more solutions to that problem. There that you learn to think mathematically.

Many of the problems in this book can be solved in more than one way. There is always a single answer, but there can be many paths to that answer. Once you solve a problem, go back and see if you can solve it by another method. See how many methods you can find. Then check our solutions to see if any of them differ from yours. *Most veteran problem-solvers agree that the solutions usually teach much more than the questions do.*

Each solution in this book is the result of *hours* of consideration and discussion. This book usually omits common, brute-force ways such as guess-check-and-revise or make-a-list whenever thoughtful, conceptual approaches are available. However, if you don't see a mathematically efficient solution, try any method that promises to work. Even if your approach to a problem, under time pressure, takes the same path as ours, there still may be much that you can learn from ours. We tried to polish each solution in this book to be direct, concise, and elegant.

There is a saying: "All of us together are smarter than any of us alone." Attempting a problem as part of a study group enables you to benefit not only from your insights but also to learn from others. While you may want to test yourself as an individual, consider working with others on problems or, alternatively, comparing solutions aloud after each of you worked separately on the same questions.

2.   Why We Study Problem Solving

Most people, including children, love puzzles and games. It is fun to test ourselves against challenges. The continuing popularity of crossword, jigsaw, and sudoku puzzles, as well as of board, card, and video games, attests to this facet of human nature.

Problem solving builds on this foundation. A good problem is engaging in both senses of the word. Child or adult, we readily accept the challenge, wanting to prove to ourselves that "I can do this." To many of us, a problem is fun more than it is work. A good problem captures our interest, and once you have our interest, you have our intelligence.

A good problem contains within it the promise of the thrill of discovery, that magic "Aha!" moment. It promises that deep satisfaction, if we solve it, of knowing we've accomplished something. It promises growth, the realization that we will know more today than we did yesterday. It speaks to our universal desire for mastery. Babies reach continually for their toes until they succeed in grasping them. Toddlers fall continually until they succeed in walking. Children swing continually at a baseball until they succeed in hitting it frequently. A good problem promises us many things, all of them worthwhile.

The history of mathematics is a history of problem-solving efforts.

It is said that perhaps all mathematics evolved as the result of problem solving. Often, one person challenged another, or perhaps himself, to handle an unexpected question. Sometimes the solution extended the range of knowledge in a well-known field or even led to the creation of a new field.

Very little of the knowledge we have today is likely to have developed in the way we study it. What we see in many books is the distilled material, whose elegant, economical organization and presentation obscure the struggling and false starts that went into polishing it. That which grabbed and held the imagination has been scrubbed out.

A typical example of this is high school geometry. The ancient Greek mathematicians enjoyed challenging each other. "Alpha" would ask "Beta" to find a way to perform a specific construction using only a straightedge and a pair of collapsible compasses. But it wasn't enough for Beta to merely create a method. He also had to convince Alpha that his method would always work. Thus, proofs were born, as a result of problem solving. Slowly, the Greeks built up a whole body of knowledge. Euclid's genius was in collecting all the properties into an organized reference book in which most of them grew logically from a very few basic properties.

For centuries we have taught geometry according to Euclid, instead of tracing the journey taken to arrive at the knowledge. This removed the thought-provoking elements of problem solving mentioned above, and with it, much of the involvement for many students.

Therefore … why should we learn math through problem solving? To put the magic back into math.

3.   Characteristics of Good Problems

Problems and exercises are quite different from each other. An exercise presents a familiar situation to which we apply a known procedure. A problem presents an unfamiliar situation for which we need to devise a procedure. For most readers, few of the problems in this book are exercises, if any. A problem is said to be good if it contains some or all of the following ten characteristics.

1.   A good problem is engaging and interesting to students. They want to see "how it ends."

2.   A good problem illustrates useful principles. The mathematics should not only *be* meaningful, but also should *appear* meaningful to the solver. It should involve important topics, whether past, present, or future. Puzzles are fun, but they may or may not fit this criterion.

3.   A good problem allows creative or subtle solutions. Some unexpected solutions may prove startling.

4.   A good problem allows a fresh look at situation. Familiar skills and principles may be used in unfamiliar ways.

5.   A good problem can trigger more than one method of solution. It is not unusual to solve one three or four different ways. For example, problem 3E on page 31 was solved by one group of fourth and fifth grade students 11 different ways![1]

6.   A good problem can combine more than one topic. It is not simple and it allows the student to see a connection between two apparently different topics.

7.   A good problem can be expanded or extended. Often, the situation presented in a problem turns out to be part of a much larger picture. A concept may be seen as a special case of a more general concept. Think of it as a camera zooming out to reveal unexpected additional things.

8.   A good problem is related to other problems. Two apparently different problems may turn out to have many elements in common.

9.   A good problem is challenging, but not intimidating, and should appear accessible to everyone. It should not demand specialized knowledge not possessed by the solver. It is more important that we develop the desire to continue to try problems.

10.   A good problem is clear to all students after explanation. This provides both closure to the problem and a sense of satisfaction, even to those who did not solve it correctly on their own.

---

[1] Kalman, R. "Revisiting the Sum of Odd Natural Numbers", *Mathematics Teaching in the Middle School* (NCTM), (September 2003: 58-61).

4.   What Every Young Mathlete Should Know

I. VOCABULARY AND LANGUAGE

The following explains, defines, or lists some of the terms in **bold** that may be used in Olympiad problems. An answer is acceptable on an Olympiad contest only if it is consistent with both this section and the wording of the related problem.

1.   Basic Terms

Sum, difference, product, quotient, remainder, ratio, square of a number (also, **perfect square**), **factors of a number.** The **value** of a number is the simplest name for that number. "**Or**" is inclusive: "*a* or *b*" means "*a* or *b* or both."
⟹ DIVISION M: **Square root** of a number, **cube** of a number (also, **perfect cube**).

2.   Reading Sums

An ellipsis (…) means "and so on":
Read "$1 + 2 + 3 + …$"   as "one plus two plus three and so on, without end."
Read "$1 + 2 + 3 + … + 10$"   as "one plus two plus three and so on up to ten."

3.   Writing Whole Numbers

The **standard form of a number** refers to the form in which we usually write numbers (also called Hindu-Arabic numerals or positional notation).

A **digit** is any one of the ten numerals 0, 1, 2, 3, 4, 5, 6, 7, 8, 9. All numerals are written by assigning place values to combinations of these digits. A number may be described by the number of digits it contains: 358 is a three-digit number. The "**lead-digit**" (leftmost digit) of a number is not counted as a digit if it is 0: i.e., 0358 is a three-digit number. **Terminal zeroes** of a number are the zeroes to the right of the last nonzero digit: i.e., 30,500 has two terminal zeroes because to the right of the digit 5 there are two zeroes.

4.   Sets of Numbers

a. **Whole Numbers** = {0, 1, 2, 3, …}.
b. **Counting Numbers** = {1, 2, 3, …}.
Note: The term natural numbers will not be used in Olympiad contests because in some fields of mathematics, the definition of natural numbers is the same as counting numbers while in other fields it is the same as whole numbers.

c. ⟹ DIVISION M: **Integers** = {…, –3, –2, –1, 0, +1, +2, +3, …};
Positive numbers, negative numbers, nonnegative numbers, and nonpositive numbers.

---

d. **Consecutive Numbers** are counting numbers that differ by 1, such as 83, 84, 85, 86, and 87. **Consecutive Even Numbers** are multiples of 2 that differ by 2, such as 36, 38, 40, and 42. **Consecutive Odd Numbers** are nonmultiples of 2 that differ by 2, such as 57, 59, 61, and 63.

5. Multiples, Divisibility, and Factors

The product of any two whole numbers is called a **multiple** of each of the whole numbers. Zero is considered a multiple of every whole number. *Examples*: The multiples of 6 = {0, 6, 12, 18, …}. *Note: many but not all authorities expand the definition of multiples to include all integers. To them, −24 is a multiple of 6. For Olympiad problems, no multiples will be negative.*

A whole number, *a*, is said to be **divisible** by a natural number, *b*, if the remainder is zero upon division. In this case: (1) their quotient is also a whole number, (2) *b* is called a factor of *a*, and (3) *a* is called a multiple of *b*.

6. Number Theory

a. A **prime number** (also, **prime**) is a counting number which has exactly *two* different factors, namely the number itself and the number 1. *Examples*: 2, 3, 5, 7, 11, 13, …

b. A **composite number** is a counting number which has *at least three* different factors, namely number itself, the number 1, and at least one other factor. *Examples*: 4, 6, 8, 9, 10, 12, …

c. The number 1 is neither prime nor composite since it has *exactly one* factor, namely the number 1. Thus, there are 3 separate categories of counting numbers: prime, composite, and the number 1.

d. A number is **factored completely** when it is expressed as a product of only prime numbers. *Example*: $144 = 2 \times 2 \times 2 \times 2 \times 3 \times 3$. It may also be written as $144 = 2^4 \times 3^2$.

e. The **Greatest Common Factor (GCF)** of two counting numbers is the largest counting number that divides each of the two given numbers with zero remainder. *Example*: GCF (12,18) = 6.

f. If the GCF of two numbers is 1, then we say the numbers are **relatively prime** or **co-prime**.

g. The **Least Common Multiple (LCM)** of two counting numbers is the smallest number that each of the given numbers divides with zero remainder. *Example*: LCM (12,18) = 36.

h. **Order of Operations**. When computing the value of expressions involving two or more operations, the following priorities must be observed from left to right:

1) do operations in parentheses, braces, or brackets first, working from the inside out,
2) do multiplication and division from left to right, and then
3) do addition and subtraction from left to right.

$$\begin{aligned} \textit{Example:} \quad & 3 + 4 \times 5 - 8 \div (9 - 7) \\ & = 3 + 4 \times 5 - 8 \div \quad 2 \\ & = 3 + \quad 20 \quad - \quad 4 \\ & = 19 \end{aligned}$$

7. Fractions

    a. A **common** (or **simple**) **fraction** is a fraction (of the form $\frac{a}{b}$) where the **numerator** and **denominator** are whole numbers, except that the denominator cannot be 0. Its operational meaning is that the numerator is divided by the denominator.

    b. A **unit fraction** is a common fraction with numerator 1.

    c. A **proper fraction** is a common fraction in which the numerator is less than the denominator. Its value is more than 0 and less than 1.

    d. An **improper fraction** is a common fraction in which the numerator is equal to or more than the denominator. Its value is 1 or greater than 1. A fraction whose denominator is 1 is equivalent to an integer.

    e. A **complex fraction** is a fraction whose numerator or denominator contains a fraction.
Examples: $\dfrac{\frac{2}{3}}{5}$, $\dfrac{2}{\frac{3}{5}}$, $\dfrac{\frac{2}{3}}{\frac{5}{7}}$, $\dfrac{2+\frac{3}{5}}{5-\frac{1}{2}}$

Complex fractions are often simplified by using the operational meaning.

Example: $\dfrac{\frac{2}{3}}{5} = \frac{2}{3} \div 5 = \frac{2}{3} \times \frac{1}{5} = \frac{1}{15}$.

    f. The fraction is **simplified** ("**in lowest terms**") if a and b have no common factor other than 1 [GCF(a,b) = 1].

    g. A **decimal** or **decimal fraction** is a fraction whose denominator is a power of ten. The decimal is written using decimal point notation. *Examples*: $0.07 = \frac{7}{100}$, $0.153 = \frac{153}{1000}$, $6.4 = 6\frac{4}{10}$ or $\frac{64}{10}$

    h. ⟹ DIVISION M: A **percent** or **percent fraction** is a fraction whose denominator is 100. The percent sign represents the division by 100. *Examples*: $9\% = \frac{9}{100}$, $125\% = \frac{125}{100}$, $0.3\% = \frac{0.3}{100}$ or $\frac{3}{1000}$

8. Statistics and Probability

The **average** (**arithmetic mean**) of a set of N numbers is the sum of all N numbers divided by N. The **mode** of a set of numbers is the number listed most often. A set with every number listed once is said to have no mode. The **median** of an ordered set of numbers is the middle number if N is odd or it is the mean of the two middle numbers if N is even.

The **probability of an event** is a value between 0 and 1 inclusive that expresses how likely an event is to occur. It is often found by dividing the number of times an event does occur by the total number of times the event can possibly occur. *Example*: The probability of rolling an odd number on a die is $\frac{3}{6}$ or $\frac{1}{2}$. Either $\frac{3}{6}$ or $\frac{1}{2}$ will be accepted as a correct probability on an Olympiad contest.

9.   <u>Geometry</u>

    a. Angles: **degree-measure, vertex, congruent; acute, right, obtuse, straight, reflex.**

    b. **Congruent segments** are two **line segments** of equal length.

    c. **Polygons and circles:**

        i. Parts: **side, angle, vertex, diagonal; interior region, exterior region; diameter, radius, chord.**

        ii. **Triangles: acute, right, obtuse; scalene, isosceles, equilateral.**
        Note: all equilateral triangles are isosceles, but only some isosceles triangles are equilateral.

        iii. **Quadrilaterals: parallelogram, rectangle, square, trapezoid, rhombus.**
        Note: a square is a rectangle with all sides congruent. It is also a rhombus with all angles congruent.

        iv. Other **polygons: pentagon, hexagon, octagon, decagon, dodecagon, icosagon.**

        v. **Perimeter**: the number of unit lengths in the boundary of a **plane figure.**

        vi. **Area**: the number of **unit squares** (also, **one-unit squares**) contained in the interior of a region. A unit square is a square each of whose sides measures 1 unit. The area of each unit square is **one square unit**. *Example*: The area of a **5-cm square** is 25 sq cm.

        vii. **Circumference**: the perimeter of a circular region.

        *viii.***Congruent figures**: two or more plane figures whose **corresponding** pairs of sides are congruent and whose **corresponding** pairs of angles are congruent.

        ix. **Similar figures**: two or more plane figures whose size may be different but whose shape is the same. *Note: all squares are similar; all circles are similar.*

    d. **Geometric Solids:**

        i. **Cube, rectangular solid; face, edge.**
        ⟹ **DIVISION M: cylinder** (right circular only).

        ii. ⟹ **DIVISION M: volume**: the number of **unit cubes** contained in the interior of a solid. A **one-unit cube** is a cube each of whose edges measures 1 unit. The volume of each 1-unit cube is **1 cubic unit**. *Example*: The volume of a **5-km cube** is 125 cu km.

        iii. ⟹ **DIVISION M: surface area**: the sum of the areas of all the faces of a geometric solid.
        Example: The surface area of a **5-cm cube** is 150 square cm.

II.   SKILLS

1.   Computation

The tools of arithmetic are needed for problem solving. Competency in the basic operations on whole numbers, fractions, and decimals is essential for success in problem solving at all levels.
⟹ **DIVISION M:** Competency in basic operations on signed numbers should be developed.

2.   Answers to contest questions

Unless otherwise specified in a problem, equivalent numbers or expressions should be accepted. For example, $3\frac{1}{2}$, $\frac{7}{2}$, and 3.5 are equivalent.

Units of measure are rarely required in answers — but if stated in an answer, they must be correct. More generally, an answer in which any part is incorrect is not acceptable. To avoid being denied credit students should be careful to include *only* required information. While an answer that differs from the official one can be appealed, credit can be granted only if the wording of the problem allows for an alternate interpretation or a flaw prevents any answer from satisfying all conditions.

Measures of area are usually written as square units, sq. units, or units^2. For example, square centimeters may be abbreviated as sq cm, or cm × cm, or cm^2.
⟹ **DIVISION M:** cubic measures are treated in a like manner.

On a contest, any answer different from the "official" answer should be marked as incorrect. However, some of these answers can be appealed. The panel of judges can grant credit only if the wording of the problem allows for a alternate interpretation or if it is flawed so that no answer satisfies all conditions of the problem.

After reading a problem, a wise procedure is to indicate the nature of the answer at the bottom of a worksheet before starting the work necessary for solution. Examples: "A = ___, B = ___"; "The largest number is ___"; "The perimeter is ___ sq cm". Another worthwhile device in practice sessions is to require the student to write the answer in a simple declarative sentence *using the wording of the question itself*. Example: "The average speed is 54 miles per hour." This device usually causes the student to reread the problem and answer the question actually asked, rather than the question he or she wishes were asked.

3.   Measurement

The student should be familiar with units of measurement for time, length, area, and weight (and for **DIVISION M**, volume) in both English and metric systems. Within a system of measurement, the student should be able to convert from one unit to another.

Because there was no year 0, the first century spanned the years 1 to 100 inclusive. Accordingly, the 20th century spanned the years 1901 to 2000 and the 21st century spans the years 2001 to 2100, all inclusive.

5.   Some Common Strategies for Solving Problems

The most commonly used general strategies include the following:

1. **Guess-check-and-revise.** This is the most basic of methods. Rarely is it quick or the one that the problem writer had in mind, but it can help you find the answer. Still, even if you get the right answer, you should ask other people or check the "official solution" to understand the mathematics behind the problem. After all, the long-term goal is to improve mathematically, not merely to score points.

2. **Find a pattern.** Mathematics is sometimes described as the science of patterns. Studying how the numbers behave in a given problem can allow you to predict the result.

3. **Make an organized list.** Listing every possibility in an organized way is an important tool. How you organize your data often reveals additional information.

4. **Build a table.** This is a special case of making an organized list. Seeing the numbers laid out in an organized way allows you to comprehend the patterns within.

5. **Solve a simpler related problem.** Many hard problems are merely easy ones that have been extended to larger numbers. Replacing the large number by at least three of the smallest numbers possible can introduce patterns that allow you to solve the original problem. Strategies 2 through 5 are often used together in various combinations.

6. **Eliminate all but one possibility (process of elimination).** Deciding what a quantity is *not* can narrow the field to a very few possibilities. Test each remaining possibility against all conditions of the problem.

7. **Handle one condition at a time.** Choosing just one of several conditions stated within a problem allows you to start with a list of possible results and then reduce its size. It is related to strategies 5 and 6.

8. **Work backward.** If a problem describes a procedure and then specifies the final result, this method usually makes the problem much easier to solve.

9. **Draw a picture or diagram.** One picture is worth a thousand words. In a geometry problem, always draw a diagram. In a non-geometry problem, try to represent the amounts by a diagram; it may make the solution much easier to understand. Label it with all given measurements and see what other numbers you can figure out.

10. **Draw a convenient segment in a geometric figure.** Sometimes you can divide an unusual shape into two or more common shapes by drawing one or more convenient line segments. Other times you can transfer a known length to a much more usable location.

---

11. **Make a change and compensate.** There are times when changing some of the conditions of a problem make a solution clearer or more convenient. However, to return to the original overall conditions of the problem requires either an additional change or the reversal of the first change. For an example of the former, change 45×16 to 90×8 or change $488 + 756$ to $500 + 744$. For an example of the latter, find the 50th odd number by shifting all members of the sequence 1, 3, 5, … up by 1 to find the 50th even number, 100, and then reverse the shift, getting 99.

Encourage students to guess-check-and-revise when no other method presents itself. With time and practice, more efficient strategies should start to present themselves.

Thorough discussions of these and many other useful topics may be found in *Creative Problem Solving in School Mathematics* and *Math Olympiad Contest Problems for Elementary and Middle Schools*. See the back of this book for information.

6. Selecting Problems for a Purpose

There are many ways to use this book. Some examples follow.

Beginning problem-solvers often need a little more help getting started. They may want to choose easy problems (those of high percentage) in order to get the "feel" of our problems and build up confidence, eventually building up to the harder problems (those of low percentage). It takes time for people to understand what is expected of them and to adopt a mode of thought. It is a good idea to minimize pressure by ignoring the given time limits at first. On the other hand, experienced problem solvers might take the opposite approach in order to sharpen existing skills, by dealing with the most difficult problems and trying the Follow-Ups.

As an Olympiad contest date approaches, some readers might wish to simulate the Olympiad under actual contest conditions, like a dress rehearsal. During earlier practices though, there is value in ignoring time limits and having small groups work cooperatively. This removes the pressure of time, allows the mathletes to review more concepts and to help each other, and involves everyone in the problem-solving process more fully. During later practices, adhering to time limits is advisable.

The reader may want to focus on one specific type of problem at a time, using the contest problem types on pages vii-x. This strategy for studying can be extended by combining a topic from *Creative Problem Solving in School Mathematics* (MOEMS, 2005) with related problems taken from this book. This permits an investigation of that topic in depth and perhaps a shoring up of a perceived weakness. With this approach, however, many interesting one-of-a-kind problems will be missed.

Each method of study mentioned above has its strengths and its limitations, but an overall plan incorporating each in turn may be the most effective preparation for the Olympiads.

7. Attempting to Solve a Problem

Solving a problem is comparable to driving with insufficient directions in a strange neighborhood. As we search for a route from point A to point B, we probe different roads hoping to find a landmark. Whenever we realize that a particular road won't get us to our destination, we go back to a previous spot and try another road. We believe that if we continue probing, we are bound to find point B. The next time we drive from A to B, we are likely to make fewer false turns, resulting in a much quicker, more efficient drive. If we continue driving from A to B, we eventually find the shortest, most efficient route.

When faced with a complex problem, we try different routes to a solution. If one approach does not work, we try another. We probe and probe, each time learning something else. We hope to find a route that takes us to an answer. If we then write out our solution or explain it orally, we eliminate many false starts, our understanding becomes fuller, and our solution becomes more direct. This is an unexpected benefit provided by rewriting a solution from scrap paper. The more frequently we explain it to other people, the deeper our comprehension and the sharper our explanation is likely to be.

A common saying is, "Problem solving is *what* you do when you don't *know* what to do."

Perhaps the most important quality that problem solvers develop is that of persistence. We believe that with each attempt, we will discover something more. On our first try we are likely to see some of the thinking behind the solution. With each subsequent attempt, we gain additional insights, encouraging us to try again. Without realizing it, this persistence gives us two major benefits: our ability to think mathematically improves, and our capacity to handle frustrations increases.

8. Solving Problems in a Contest Setting

Picture yourself in the middle of a math contest. Unfortunately, not only does the problem appear to be different from any you have seen before, but you are told you have only five minutes to solve it. *How* do you proceed?

Start by reading the problem through twice. The first time, look for the overall picture: Where does the problem start? Where does it finish? What is the general situation or type? The second time, notice as many details as possible. Can you draw any conclusions from any given fact? Can you connect any two facts together? Spending a little extra time at the start can save you much time afterwards.

It pays to focus on each fact and each number given in the problem. The author probably chose them for a reason. In fact, changing any fact or number often changes the nature of the problem. Ask yourself *how* you can use any or all of the given facts or numbers.

Next, decide on a course of action, using all or some of the following approaches.

- **Is the problem similar to a more familiar problem?** Can you apply the same general method that worked before? What changes are needed to make the method more appropriate for the current problem?

- **Work forwards from the beginning.** Can you use any fact in the problem to learn something more? Can the second thing in turn reveal a third thing? How far can you extend this chain of reasoning? This course is similar to a toddler putting one foot in front of the other when learning to walk. Call it Toddler Math.

- **Work backwards from the end.** Does the desired answer suggest a step *before* it? Is there a fact or number you must find before you can reach the concluding step? Is there a step before that which will lead you to the next to last step? How far back from the conclusion can you trace the chain of reasoning?

- **Combine the two approaches.** Many problems are solved by working from both the beginning and the end towards some middle point. At best, the two will meet. At worst, you could have a much simpler problem to solve.

- **Guess-check-and-revise.** As mentioned on page 10, this may be the most inefficient method of solving problems, but it is the easiest way to reach the correct answer. It often has the strength of revealing things about the problem with each guess and just as often the weakness of hiding the concept underlying the problem. This is the method to use if you see no other way to solve a problem; it gets you started. Beginners in particular are urged to use guess-check-and-revise often, until they catch the "feel" of solving problems. In time, as you learn to think mathematically, you grow into using more sophisticated and efficient approaches.

- **Get to know many of the most common strategies.** In Section 5, several were discussed. Become comfortable with them to allow you to see ways to handle many problems.

Every Math Olympiad problem is built to encourage mathematical thought on the part of the student. Nevertheless, for purposes of entering scores, it makes no difference how the answer was obtained, only that the students found the correct answer. Of course, to improve your problem solving abilities, examine the "official solution" after you try each of the problems.

9.   Why We *Teach* Problem Solving

"When am I ever going to use this?" Contrary to common belief, these words are rarely uttered by a student intent on discussing the future. Rather, they seem to be a statement of frustration, an indirect way of saying, "I don't know what I'm doing, and I don't like feeling uncertain." Cataloging careers that use mathematics or listing high school math courses does not seem likely to reassure such a student. Reassurance comes best from giving the student ownership of the skill or concept, and by helping the student become comfortable with the work.

Teachers can provide ownership by posing a good problem and then by allowing the student to find his or her own way to solve it. Teachers can use that innate love of a challenge to engage the student's interest, and therefore the student's intelligence, as discussed in Section 2. The student, responding to the challenge, sees reasons to focus on the problem. If the problem is well chosen and constructed, the student learns something important.

How does teaching through problem solving help the teacher to develop stronger students?

1. **Interest:** Good problems tap into something deep within all of us. Assigning them to students engages their interest and focus.

2. **Meaning:** Embedding a concept or skill into a good problem provides more meaning to students than merely stating or demonstrating it; students can see a purpose. Skills are not learned just for their own sake, but are seen as tools to be used, not as ends in themselves.

3. **Complexity:** Life is not simple. Life's problems are not simple. We do students no favor by making mathematical problems simple. Most good problems require more than one concept to solve. Assigning good problems may help students develop the ability to see the interplay between different concepts.

4. **Creativity and flexibility:** Many of the problems in this book show multiple solutions. The more ways a student can see to solve a problem, the freer he or she may feel to try an unusual approach to a puzzling problem. Some of the problems in this book have triggered highly inventive methods.

5. **Developing mathematical thinking:** Over time, continued exposure to thoughtful solutions leads students to think mathematically. Subsequent math courses in high school will require this ability, as will college entrance examinations.

6. **Retention:** A good problem often allows for many strategies and may require several principles and skills with each strategy. Continually tackling problems and discussing solutions allows the student to revisit most concepts and procedures frequently, each time from a fresh perspective. This disguised practice builds in reinforcement while it clarifies the concept.

*<u>P</u>erson <u>In</u> <u>C</u>harge of <u>O</u>lympiad

7. **Building student confidence:** Each demanding problem the teacher assigns reflects a respect for the students' abilities, which students may well appreciate. With each successful solution, the student realizes, "I can do this!" Since nothing builds self-confidence like accomplishment, these two things can stimulate intellectual growth.

8. **Empowerment:** When the teacher allows students to tackle problems on their own with no more than an occasional hint and then asks several students to present different approaches aloud, the students assume ownership of the problems. Students are likely to accept that the responsibility for learning is theirs, not the teacher's. Each student becomes an active partner in the learning process. The problem, the solution, the principles involved all become his or her property and are more likely to be available when needed.

10. Why Should We Do MOEMS Contests?

As discussed in Section 2, various puzzles, games, and sports leagues are immensely popular among both children and their parents, and have been so for many generations. In fact, children seem to turn almost any situation into a game. Why is this so? The most basic answer is that we love to compete, to test ourselves against arbitrary standards, our potential, or our peers. All we require is a reasonable chance of succeeding. Why do we keep at it if we are not doing well? We have an innate belief that with practice we will continually improve. In most of the above we compete less against other people than against our own abilities. This is a valuable source of true growth.

Contests such as spelling bees and the Math Olympiads use this fundamental human characteristic to entice students into mastering skills or developing a way of thinking. Children (and sometimes the adults) may assume they are competing in order to win awards or to show they can do well on a national scale. The real purpose, however, is to get them to want to tackle richer, more demanding problems than those they may be used to. As with all education, the goal is growth. Adults should be careful to keep the students' sense of competition low-key. To maximize growth, the children should not feel pressured or threatened.

Students usually want two things out of any activity: to feel they belong on this level and to maintain a sense of progress.

To fill the first need and to satisfy most students, the Math Olympiad contests are scaled so that the average student correctly solves about 40% of the problems during the year. Each contest includes problems specifically designed for both ends of the spectrum. Some problems are uncomplicated (even though they involve mathematical thinking) and their solutions can easily be understood by all. Such problems enable beginners to expect to solve their fair share of the problems within the time limits. Moreover, a student's scoring typically improves from year to year. Meanwhile, other problems challenge even the strongest mathletes. Success is quite attainable, but perfection is rare. In fact, less than one-third of one per cent of all participants correctly solve every problem for the year!

To fill the second need, all Olympiad problems involve higher order thought. Those who use guess-check-and-revise will be shown more elegant and efficient solutions when the problem is reviewed immediately afterwards. Even the best mathletes will learn new ways to apply familiar concepts inasmuch as the 25 problems may employ 40 or more principles, some in very subtle ways. The variety itself presents a challenge. Each time students learn something new, their sense of accomplishment is reinforced.

There are two completely separate divisions, **DIVISION E** (for students in grade 6 and below) and **DIVISION M** (for students in grade 8 and below). The Math Olympiads consist of five monthly contests, stretching from November to March. This provides constant reinforcement of mathematical thinking and problem solving, and helps to keep student interest high. Contests are conducted under strict standard testing conditions. All team members take the contests together. There are no make-ups. Each team can have as many as 35 students and a school can have as many teams as it wants; interested students should not be turned away.

Practices are important. Beginners develop a comfort level with contest problems and all mathletes improve their problem solving skills. Towards this end MOEMS provides 50 practice problems with detailed solutions prior to the first Olympiad. Each solution also names the strategy employed; about half of the problems offer extensions to help the teacher develop each problem into a mini-lesson.

An individual's score is just the number of problems answered correctly. While in general work should be shown, it is not required on our contests. The team score, compiled after the last Olympiad, is simply the sum of the ten highest individual scores. Large teams and small teams are therefore on a fairly equal footing. No traveling is required; contests are held in the school.

Further, the generous awards structure enables the school to provide ample recognition. The awards package sent to each team includes a Certificate of Participation for every mathlete reported, embroidered felt patches for about 50% of all mathletes nationally, a variety of other awards for the high scorers on each team, and assorted team awards for about 25% of all teams. Further information is available at our Website _www.moems.org_.

11. Building a Program

So you want to start a math team.

Any successful program requires solid support from many elements. Your students must be enthusiastic about the activity, their parents must see potential benefits for their children, and the school administration must feel they can justify the expenditure easily to central administration and others. Experienced Math Olympiad coaches have used the following suggestions successfully. Few people do everything mentioned below, but the more you do, the more effective you are likely to be.

One way to build a program is to meet with the school principal and the director of mathematics to map out your procedures for devising a philosophy, recruiting students, building parental support, training students, conducting contests, and recognizing accomplishments.

Devising a Philosophy: Many decisions need to be made, including the following:

- Will the activity be open to all students or only to those in a gifted program?
- Will it be embedded in classes, done as a pull-out program, or offered as a club that meets outside of class time?
- Will it be run by a faculty member or by one or more parents?
- Will funding come from the school, from the parents' association, or from another source?
- Will students be given study and practice sheets as needed or will contest books be issued to all members?
- Will homework and outside projects be assigned?

Recruiting Students: If participation in the contests is to be voluntary, you, the PICO, should devise a campaign to encourage students to join the new math team. You may speak to several classes, ask your colleagues to talk it up in class, place posters around the school, send announcements home, write one or more articles for the school district newsletter or a local newspaper, and/or make a brief morning announcement for several consecutive days ("Only 4 more days until …"). You might ask fellow teachers to recommend names of students. Approaching one person at a time face-to-face is usually very effective. You might even sit down with the student and parent together. If you call a meeting for all interested students and their parents, be sure to leave time for them to ask you questions.

Involving students: There is value in assigning productive roles to your mathletes. You may want to hold elections for your math team officers, assigning specific duties to each office. The president and vice-president, and maybe others, might prepare and run some of the practices, with guidance from you. A team publicist might write bulletins for the school's morning announcements after each contest and perhaps also for the community newspaper or district newsletter. A recording secretary might keep track of attendance, completed assignments, and results of each contest and practice problems. Other possible student jobs include decorating bulletin boards with team news and announcements, providing snacks at practices, and scheduling social get-togethers. Holding elections for several productive positions can supply students with a vested interest in the success of the team. Your mathletes will own the activity.

Building parental Support: Parental support helps build a popular ongoing program. The better informed that parents are, the more their enthusiasm is felt by their children. Like Johnny Appleseed, parents of current mathletes are likely to "spread the word" to parents of future mathletes. Think of it as having a score of publicists working for you free. Some may volunteer to write articles for local newspapers after each contest and after special occasions, such as elections of officers and award ceremonies. In some schools parents are used to keep records and submit contest results, to open their homes to all team members for periodic social events such as barbecues, and to conduct team outings to such as ballparks or beaches. In certain cases, the parents' association or individual parents have supplied the PICO and/or paid the enrollment fee.

The next few sections discuss ways to conduct practices and contests and to recognize student accomplishments.

12. Conducting Practice Sessions

What does a practice session look like?

You will have to decide how you will choose problems, how you will configure your mathletes, and how you will utilize the students. Some of this will arise from the nature of rich problems. Mathletes do not practice a single action repetitively, because almost every problem differs from those before it. Instead, they typically need to apply more than one skill and more than one concept sequentially as they attempt to handle new and novel sets of conditions. Often they have to use familiar tools in a completely unexpected way. Your teaching will need to be geared to developing problem solving skills, not procedures. Naturally, you will personalize the routines utilized during practice, selecting appealing features from others, and adding touches of your own.

Problem selection: Your team may spend an hour investigating a variety of problems that utilize a single concept in sharply different settings. Or it may spend the time exploring a set of completely unrelated problems. The former provides depth for one concept while the latter allows for broadness, embracing many one-of-a-kind situations. Problems are assigned, either singly or in sets. You might select the problems and run the practices yourself or you might have individual team members do it. There is much of value to the latter approach.

Time and homework: We generally recommend practice sessions of at least an hour a week, but naturally, more is better. You can maximize your efforts by assigning problem sets to do at home between practices. Make sure your mathletes know that they are expected to solve some of the problems, but not all. Homework, even about problem solving, has several benefits. It extends significantly the amount of practice for each student without increasing your workload; it communicates to the student how serious you are about learning; it allows parents to see what is expected of their children; and it shows respect to the student by implying that you think he or she can handle difficult work.

Organizing your team: You can treat your team as a whole or you can split it into mini-groups. You can employ the format of a formal lesson or you can have several independent study groups work simultaneously.

Some of the practices may take the form of small in-school contests, whether team or individual, with small prizes such as candies or stickers offered as awards. Some problems may be solved by using manipulatives or acting out, and others by guess-check-and-revise or by thought-filled reasoning. You may wish to use several of these techniques.

Trusting students: One approach that has worked well is to assign a set of problems as homework, giving the students the problems *and* the detailed solutions. This allows the student to practice problem solving more frequently, which in turn can help them develop the quality of thinking mathematically. Suggest that they attempt one or two contest problems every day, adhering to time limits and test conditions.

The student tries the problems, and then, using the printed solutions, determines the number of problems solved correctly. At the next practice you record their numbers. This keeps them on task. Most students are likely to hold themselves to higher standards than anyone else could. With solutions available, many will opt to see how MOEMS handled a troublesome problem; suggest it yourself if students don't think of it. Others, even though they solved the problem, may look at our solution to verify their approach or to see if our solution contains anything unexpected.

Cooperative solving: Form your team into small groups of three or four mathletes and let them discuss the problem. Problem solving often involves discovering the correct solution a piece at a time; it requires getting past false starts and turns. A given or implied fact is overlooked at first, or a small careless error changes the nature of the question. Discussions in a group allow members to continually correct and teach each other, moving ever closer to a complete solution. Circulate, monitoring each group's discussion continuously. Have some, occasionally all, groups share their discussions with the whole team.

Multiple solutions: Many problems can be solved two or more different ways. A powerful technique is to allow several mathletes to explain several different methods of solution. After each explanation, ask if anyone used a different approach. After everyone has had the opportunity to speak, then you can add, "You know, there's still another way." By explaining more methods aloud, more concepts are reviewed and more students receive recognition.

Giving students ownership: As stated above, let the students explain their solutions aloud. Let teammates correct errors, where possible. Do not offer your solution unless it provides an insight not stated by the mathletes. There are many reasons why it is more effective for several youngsters to explain solutions than for the teacher to do so: Most people remember what they say better than what they hear; students listen more closely to friends than adults; when explaining, a student revisits a problem, often seeing aspects not noticed before; students recognize that the teacher has shown them the respect of asking them what they think. At first this practice of allowing many students to comment on a problem is much slower than lecturing; you cover less material. But as students take ownership of the learning process, their ability to think mathematically increases, resulting in a faster learning process. By the end of the year, your team will have covered more material than if you had lectured, and they would have done so more thoroughly.

13. Conducting Practices after Each Contest

"Practice makes perfect." As a rule, the more frequently your mathletes practice, the more they learn and the more capable they become. Cramming for a once-a-year test or contest helps to some extent, but not nearly as much as preparing for it on a regular basis. In fact, *working for a few minutes every day is more effective than working for several hours periodically*. Learning is reinforced while it is fresh and forgetting is minimized. That said, most mathletes practice for an hour every week, as a compromise between the ideal and other demands. PICOs can supplement the formal practice, though, by assigning homework.

It is best to review solutions at the time of the contest itself, while the problems are freshest in the mathletes' minds. The first practice after a contest can be used to extend some contest problems. You might combine *Follow-Up* problems chosen from the contest with other related problems selected from this or other books. It is recommended that, when possible, you try to choose problems that involve the same concept as the contest problem. These should use the concept in a different way or combine it with another principle.

There are two ways to go with practices, in general. You might want to cover a wide range of topics, concepts, and types. This book and *Math Olympiad Contest Problems for Elementary and Middle Schools*, which contains another 400 problems with detailed solutions, are ideal sources, offering complete contests. Or you might want to focus on fewer concepts, examining each in depth. *Creative Problem Solving in School Mathematics* offers approximately 400 more problems, arranged by topics. Many PICOs prefer to mix the two approaches.

14. Topping Off the Year

Once the last contest is done, two things remain to do: distribute awards and set up the team for the following year.

Distributing awards: Your mathletes have worked hard and have grown significantly. They have earned recognition. Towards this end, many teams create a special awards assembly. It may occur during school, after school, in the evening, or on a weekend. It may be attended by the entire student body, by parents and guests, or by team members only. It may take place in the school auditorium, in a classroom, or at a team barbecue or dinner. What is important is that students are honored publicly.

A special school-wide assembly or an evening ceremony might contain any of the following: Parents and a guest speaker can be invited. The speaker might be a member of the board of education, the district superintendent, the high school chairman of mathematics, or a professor of mathematics from a local college. This can help the students see how this year's activities can lead to future activities or how they fit into a larger picture. It also acknowledges the importance of the activity. During the ceremony the PICO might alternate past Olympiad problems with the awards and elicit solutions aloud from the mathletes. This provides a public way to acknowledge what the students can do.

Setting up the team for the following year: Make sure each officer's responsibilities are clear and well understood. Increase the vested interest each mathlete has in the new team by having them elect the new officers. This would give the officers time to plan ahead. Also, by holding the vote at this point, you limit the officers to those with at least one year's experience as mathletes.

Training the returning members of next year's team: The school year contains several weeks after the "season" is over. They can be used to sharpen the problem solving abilities of your "veterans-to-be" and perhaps to go into selected topics more deeply. After several months of straightforward practices, a change of pace can be very effective. Messages from PICOs have described some ingenious activities. You may want next year's leaders to actually plan and teach the lessons. A lot can be learned just from the act of choosing problems to fit a plan. Rotating the responsibilities among many students strengthens more than just one or two of them.

Team competitions: Split the team into small, nearly equal groups. Give them a sheet of 5 problems and a time limit of 10 or 15 minutes and have them work cooperatively on the set. A tight time limit may force them to each tackle a different problem, and when done, tackle another. Give them one point for every correct answer and a bonus point if they answer all five problems correctly. Keep a running total for a few sessions.

Scavenger hunt: Starting in one room, the scavenger hunt quickly spreads to many parts of the building. The team is split into several small groups. Past problems are chosen or rewritten so that the answer indicates a specific location in the school. As a group solves a problem, it goes to the location indicated by the answer. The person posted there accepts the answer and gives the group a new problem, which can lead the group to another location. This continues until the last problem, which leads the group to the finish point. The winner is the first group to get to the finish point.

Relay race: Again, the team is split into small groups. A problem is distributed to each group, which then shows its answer to the PICO. If the answer is correct, the group receives the next problem. Each correct answer allows the group to receive the following problem. The first one to solve the last question correctly wins. A variation is for the PICO to accept answers regardless of correctness. In this case, the team with the most correct answers wins.

Tournament: The current team might participate in a three-way contest. A second team can be formed from the parents of the mathletes and a third team from past members of the team. Holding the contest in the evening or on a weekend allows many people to attend. This also allows parents and friends to witness "their mathlete" in action, and adds to the excitement.

Alternately, your local teacher organization or mid-sized school district might host a full-day face-to-face tournament for five-member teams from several schools. Consisting of more than one event, the tournament would include problems like those found in this book, solutions-review sessions, and presentation of awards. For a moderate fee, MOEMS would supply camera-ready problems and solutions, and a detailed booklet walking your tournament committee through every step from preparation to execution. The latter includes camera-ready masters for all paperwork and a computer-ready scoring spreadsheet. Your organization or district would set and keep the per-team fees and would control every step of the way.

Any of these events provides a cap to the year, allowing the students to finish on a high note. If you've successfully employed activities not expressed in this book, please contact us at *office@moems.org* with a detailed description. We would *love* to know about it and perhaps share it with our PICOs.

OLYMPIAD PROBLEMS

DIVISION E
SETS 1-10

1A.
3
Minutes

What is the value of the following, in simplest terms?

$$(20 \times 24 \times 28 \times 32) \div (10 \times 12 \times 14 \times 16)$$

16

1B.
5
Minutes

Roni starts with the number 5 and counts by 8s. This results in the sequence 5, 13, 21, 29, 37, and so on. What is the twenty-fifth number in the sequence?

197

1C.
6
Minutes

A line segment (such as $\overline{ED}$ as shown) that connects any two points of a circle is called a *chord of the circle*. How many different chords, including $\overline{ED}$, can be drawn using only points *A*, *B*, *C*, *D*, and *E*? (*Note: $\overline{ED}$ is the same as $\overline{DE}$*)

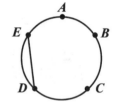

10 chords

1D.
5
Minutes

A represents a counting number. Find the value of *A* if:

$$\frac{A + A}{A \times A} = \frac{1}{3}$$

A=6

1E.
6
Minutes

Ben and Jerry start with the same number of trading cards. After Ben gives 12 of his cards to Jerry, Jerry then has two times as many cards as Ben does. How many cards did Ben have at the start?

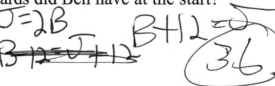

B+12=2B
12/B=1 b=2y
J=2B
B-12=J+12
B+12=J
36

SET 1 OLYMPIAD 2

2A.
3 Minutes

Marty has 6 more pogs than Jen has. After he gives 10 pogs to Jen, how many more pogs will Jen have than Marty?

14

2B.
4 Minutes

A rectangular box is 2 cm high, 4 cm wide, and 6 cm deep. Michelle packs the box with cubes, each 2 cm by 2 cm by 2 cm, with no space left over. How many cubes does she fit into the box?

2 6
4

46

6 cubes

2C.
5 Minutes

At the right, boxes represent digits and different letters represent different non-zero digits. What three-digit number is the least possible product?

299

23
× 13

1 3
2 3
3 9
2 6
2 9 9

2D.
5 Minutes

List all counting numbers which leave a remainder of 4 when divided into 22.

9, 18, 6

2E.
6 Minutes

Admission to the local movie theater is \$3 for each child and \$7 for each adult. A group of 12 people pay \$64 admission. How many children are in this group?

15
49
64

$x + y = 12$
$3x + 7y = 64$

$3x + 3y = 36$
$3x + 7y = 64$
$4y = 28$
$y = 7$

Problems - Division E Olympiads

25

SET 1 OLYMPIAD 3

3A.
4
Minutes

Suppose a standard twelve-hour clock now shows a time of 10:45. What time will the clock show 100 hours from now?

2:45

3B.
5
Minutes

The tower shown at the right is made by placing congruent cubes on top of each other with no gaps. Not all cubes are visible. How many cubes does the tower contain?

5, 4, 3, 3, 2, 2, 1, 1, 1, 1

23 cubes

3C.
4
Minutes

The Panthers team won exactly 2 of its first 9 games. By winning all its remaining *N* games, the Panthers ended with victories in exactly half of the games it played. What number does *N* represent?

5

3D.
6
Minutes

If Mrs. Murphy separates her class into groups of 4 students each, 1 student is left over. If she separates her class into groups of 5 students each, 2 students are left over. What is the least number of students the class could have?

17

3E.
5
Minutes

If 16 is added to one-third of a number, the result is three times the number. What is the number?

6

4A.
4
Minutes

What is the value of the whole number N, if:

$$N = \frac{1}{2} \text{ of } \frac{2}{3} \text{ of } \frac{3}{4} \text{ of } \frac{4}{5} \text{ of } 100?$$

20 40 60 80

4B.
5
Minutes

$ABCD$ is a rectangle whose area is 12 square units. How many square units are contained in the area of trapezoid $EFBA$?

9 un^2

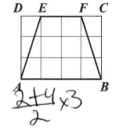

$\frac{2+4}{2} \times 3$

4C.
5
Minutes

The number 6 has exactly four different factors: 1, 2, 3, and 6. How many different factors does the number 36 have?

1, 2, 3, 6, 12, 18, 36

7

4D.
5
Minutes

A car needs 1 minute 30 seconds to travel a distance of 1 mile. At this rate, how many miles will the car travel in 1 hour?

3 min : 2 miles
60 min : x miles
x = 40

4E.
6
Minutes

At a special sale, all pens are sold at one price and all pencils at another price. If 3 pens and 2 pencils are sold for 47¢, while 2 pens and 3 pencils are sold for 38¢, what is the cost of a set of one pen and one pencil, in cents?

Pen = 12¢
Pencil = 4¢

$3x + 2y = 47$
$2x + 3y = 38$
$y = 4$ $x = 12$

$6x + 4y = 94$
$6x + 9y = 114$
$0 - 5y = -20$

SET 1   OLYMPIAD 5

5A.
3
Minutes

In the number 203,500, the last two zeroes are called *terminal zeroes*. If the multiplication $30 \times 40 \times 50 \times 60 \times 70$ is done, how many terminal zeroes will the product have?

4 terminal zeroes

$30 \times 40 = 1,200$

$1,200 + 60 = 6,000$

$36,000 \times 70 = 2,520,000$

$(30 + 6) \times 7 = 30 \times 7 + 6 \times 7$
$= 210 + 42 = 252$

$6,000 \times 60 = 36,000$

5B.
6
Minutes

Square *ACEG* is drawn at the right. Points *B*, *D*, *F*, and *H* are the midpoints of the sides of the square. What is the total number of squares of all sizes which can be traced using only the line segments shown?

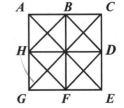

10 squares ✓

5C.
5
Minutes

Megan has *A* quarters and *B* dimes with a total value of $1.95, where *A* and *B* are both counting numbers. How many different values of *A* can Megan have?

✓

5D.
6
Minutes

On a standard circular 12-hour clock, the numerals 12 and 6 are opposite each other. On the planet Bajor, they use a circular ten-hour clock with the numerals 1 to 10 equally spaced. What pair of opposite numerals on a Bajorian clock has a sum of 11?

5E.
6
Minutes

On a shopping trip, Gil spends $\frac{1}{3}$ of his money in store *A*. Then he spends $\frac{1}{3}$ of the money he has left in store *B*. Finally, he spends his remaining $12 in store *C*. How many dollars did he have at the start?

✓ $\frac{1}{3} + \frac{1}{3} = \frac{2}{3}$ ③   *leftover of 36$*   *is 12$* $3 \times 12 = 36$ $

1A.
3 Minutes

What is the value of the following?

$$9 + 91 + 18 + 82 + 27 + 73 + 36 + 64 + 45 + 55$$

✓ 500

1B.
4 Minutes

The stairway at the right is made by placing identical cubes on top of each other. Not all cubes are visible. How many cubes does this stairway contain?

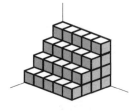

✓ 50

1C.
5 Minutes

Linda wants to buy 20 crayons. *ToyWorld* sells crayons at 4 for 25 cents, and *GameLand* sells crayons at 5 for 30 cents. Which of the two stores sells 20 crayons for less, and by how many cents less?

✓ (Gameland) 5¢ $6.25   12.5¢ TW
      20   1.20
      0580

1D.
5 Minutes

In a class of 26 students, 15 like vanilla ice cream and 16 like chocolate ice cream. However, 3 do not like either flavor. How many students like both vanilla and chocolate ice cream?

✓ 8 8 7   (8)   26-3=23

1E.
5 Minutes

The average weight of a group of children is 100 pounds. Todd, who weighs 112 pounds, then joins the group. This raises the average weight of the group to 102 pounds. How many children were in the original group?

✓ (5)   500   612/6 =

SET 2 OLYMPIAD 2

2A.
4
Minutes

Ana divides the number *N* by 8 correctly and gets .25 as her answer. Barney multiplies the same number *N* by 8. What answer should he get?

2B.
5
Minutes

Sherry has five times as many computer games as Jay. Nairan has three times as many computer games as Jay. Sherry has 16 computer games more than Nairan has. How many computer games does Jay have?

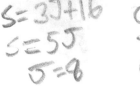
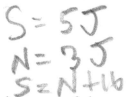

2C.
5
Minutes

Each square contains one of the numbers 1, 2, 3, 4, 5, or 6 and each number is used once. Except for the top row, the number in each square is the difference of the numbers in the two squares above it. What number is in the square marked *A*?

2D.
5
Minutes

Square *ABCD* is composed of nine congruent squares as shown. The area of the shaded region is 14 square centimeters. What is the area of square *ABCD*, in sq cm?

2E.
6
Minutes

When two people shake hands with one another, that counts as one "handshake." Every person in a room shakes hands with each other person in the room exactly once. There are a total of 15 "handshakes". How many people are in the room?

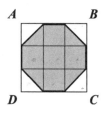

18/20

3A.
5
Minutes

N is a number such that $1 \times 2 \times 3 \times 4 \times 5 \times 6 \times 7 = 8 \times 9 \times 10 \times N$.
What is the value of N?

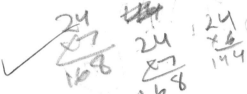

$4 \times 6 \times 7 = N \times 4 \times 3 \times 2$

$168 = 24n$

$n = 7$

3B.
3
Minutes

What number between 121 and 149 is exactly divisible by both 3 and 5?

135

3C.
5
Minutes

Rachel left home for school at 7:45 one morning. She returned home at 4:05 that afternoon. How many hours and minutes was she gone?
(The number of minutes in your answer must be less than 60.)

20 min
8 hours

3D.
6
Minutes

The area of a square is 36 square centimeters. A rectangle has the same perimeter as the square. The length of the rectangle is twice its width. What is the area of the rectangle in square cm?

32 cm² 6 cm 24

3E.
7
Minutes

Each row of ✳s has two more ✳s than the row immediately above it, as shown. Altogether, how many ✳s are contained in the first twenty rows?

and so on.

1 - 1   3 - 5
2 - 3   4 - 7   400

1-1   3-9
2-4   4-16

4A.
5
Minutes

The number 6 has exactly four different factors: 1, 2, 3, and 6. How many counting numbers less than 15 have exactly two different factors?

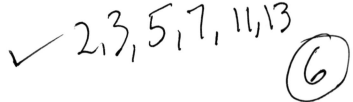

$\checkmark$ 2, 3, 5, 7, 11, 13   ⑥

4B.
5
Minutes

The scale drawing of a rectangular room measures 10 cm long by 6 cm wide. The actual width of the room is 15 feet. What is the actual perimeter of the room, in feet?

$\checkmark$ (80 ft²)   25   15

4C.
5
Minutes

2^n means that 2 is used as a factor n times. For example, 2^4 means $2 \times 2 \times 2 \times 2 = 16$. We say that 2^4 ends in 6 and that 2^4 has a units digit of 6. If 2^{1997} is multiplied out, what is the units (ones) digit of the product?

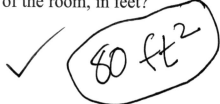

2000 6 8
99 96 4
98 97 ② $\checkmark$

4D.
5
Minutes

At a special sale, all shirts sell at one price and all caps sell at another price. Kathy pays $30 for 3 shirts and 2 caps. Marc pays $23 for 1 shirt and 5 caps. For how many dollars does each shirt sell?

$\checkmark$ ($8)

$3s + 2c = 30$
$s + 5c = 23$

$3s + 15c = 69$
$3s + 2c = 30$
$s = 8$
$13c = 39$
$c = 3$

4E.
6
Minutes

Maria writes the same whole number in each box below and gets a true statement. Jon does the same as Maria and also gets a true statement. But Jon and Maria choose different numbers. What two numbers do they choose?

$$12 + (\square \times \square) - (7 \times \square) = 0$$

$x^2 - 7x + 12 = 0$
$(x - 3)(x - 4)$

$\checkmark$ (3, 4)

5A.
3
Minutes

The figure at the right is cut out on the thick outer lines and folded on the thin inner lines to form a cube. Which letter will be on the face of the cube opposite the letter *T*?

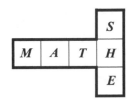

(M) ✓

5B.
5
Minutes

Trini is 1 month older than Billy. Jason is 3 months older than Billy. Kimberly is 4 months older than Jason. How much older than Trini is Kimberly, in months?

$T = B + 1$
$J = B + 3$
$K = J + 4$

$J = T + 2$

(6) ✓

5C.
6
Minutes

In this subtraction, the boxes contain the digits 3, 4, 6, and 9 in some order and the circles contain the digits 4, 5, 8, and 9 in some order. What four-digit number is represented by the boxes?

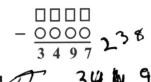

2,3,5,8
45,8,9

2,3,5,8
45,8,9

9 4589

95
98

45
58
—
7

(9346)

5D.
5
Minutes

A walkway (the shaded region) is built around a rectangular pool, as shown. The pool is 20 feet by 30 feet. The walkway is completely tiled with whole square tiles 2 feet by 2 feet. What is the fewest number of tiles that can be used?

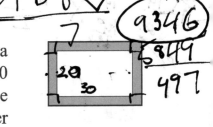

(52)   20
          30

5E.
5
Minutes

Alexandra gives $\frac{1}{2}$ of her marbles to Tyler, who gives $\frac{2}{3}$ of what he receives to Jan, who gives $\frac{3}{4}$ of what she receives to Jerry. If each has a counting number of marbles, what is the fewest number of marbles that Alexandra could have started with?

(12), 6, 4, 3

✓

1A.
3
Minutes

If $82 + 18 + 83 + 17 + 84 + 16 + 85 + 15 + 71 + N = 500$, what is the value of the counting number N?

$\checkmark$

1B.
5
Minutes

Each whole number from 1 to 9 is placed in a circle so that the sum of the three numbers along any straight line is the same as the sum along any other straight line. Which number must be in the circle marked A?

 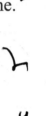

$A = 8$

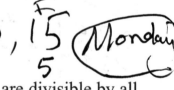

1C.
5
Minutes

Pat reads 1 page of a certain novel every Monday, 2 pages every Tuesday, 3 pages every Wednesday, and so on up to 7 pages every Sunday. She reads every page exactly once and does not skip any days. Suppose Pat starts reading page 1 on a Monday and reads the pages in order. On which day of the week does she read page 100?

$\dfrac{n^2+n}{2} = \dfrac{n(n+1)}{2}$

M T W T F
1, 3, 6, 10, 15 Monday
1 2 3 4 5

1D.
6
Minutes

In all, how many counting numbers from 200 to 500 are divisible by all of the following:

2, 3, 4, 5, and 6?

careless

9

60, 120 360 420

240, 300

1E.
6
Minutes

A circle and a triangle overlap as shown. The area of the circle is three times the area of the triangle. If the common region is removed, then the area of the rest of the circle would be 14 square centimeters more than the area of the rest of the triangle. What is the area of the complete triangle, in sq cm?

21 7

$x+y = \cancel{3z+3y}$

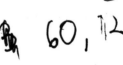

$\checkmark$

7 cm

$x+y = 3z+3y$

$x = 2+14$

Problems - Division E Olympiads

2A.
3
Minutes

What is the value of the following?

$$123 + 123 + 123 + 123 + 123 +$$
$$123 + 123 + 123 + 123 + 123 +$$
$$123 + 123 + 123 + 123 + 123 +$$
$$123 + 123 + 123 + 123 + 123$$

2460   4 8 12 16 20

123
20
000
246
1460

2B.
5
Minutes

Lee walks along the edges of a rectangular pool (shown) from point *A* to *B* to *C* to *D*, a distance of 38 meters. Marina walks along the edges of the same pool from *B* to *C* to *D* to *A*, a distance of 31 meters. What is the perimeter of the pool, in meters?

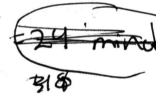

$x=8$ A $y=15$ B

D C

46 m

2x+4y=76
2x+y=31
3y=45
y=15

2y+x=38
2x+y=31

p-x=38
p-y=31

y=x+7
4x+14=

2C.
5
Minutes

Terry drives 20 miles at an average rate of 50 miles per hour while Sherry drives 20 miles at an average rate of 40 miles per hour. Neither person stops. Terry needs *M* fewer minutes to complete the trip than Sherry does. What is the value of *M*?

6 minutes

$\frac{50}{60} = \frac{40}{x} = \frac{2400}{50} = \frac{201}{3}$

$\frac{40}{60} = \frac{20}{x}$ 24

$\frac{1200}{50}$ 24 minutes

318

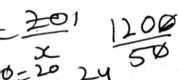

2D.
6
Minutes

At the Math Store each circle costs one amount and each square costs another amount. Five circles plus one square cost 20¢. Two circles plus three squares cost 21¢. At these prices, how many cents does twelve circles plus five squares cost?

61¢

36
25
61

$5c + s = 20$
$2c + 3s = 21$

$15c + 3s = 60$
$2c + 3s = 21$
$13c = 39$
$c = 3$ $s = 5$

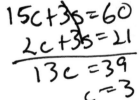

2E.
7
Minutes

Fifteen darts have landed on the dartboard shown. Each dart scores 3, 5, or 7 points. In how many different ways can the fifteen darts score a total of 75 points?

SET 3   OLYMPIAD 3

3A.
4
Minutes

Lauren is the youngest of four sisters. The average of their ages (all different whole numbers) is 9 years. What is the oldest that Lauren can be, in years?

30 = 36

~~33, 17~~ 1, 2, 3, 30

3B.
5
Minutes

The six-digit number $63X904$ is an even multiple of 27. What digit does X represent?

✓

22

$X = 5$

3C.
6
Minutes

Michelle's Number Recycling Machine obeys exactly two rules:
1. If an inserted number has exactly 1 digit, double the number.
2. If an inserted number has exactly 2 digits, compute the sum of the digits.
The first number Michelle inserts is 1. Then every answer she gets is inserted back into the machine until fifty numbers are inserted. What is the fiftieth number to be inserted?

✓

16   1, 2, 4, 8, 16, 7, 14, 5, 10, 1

3D.
6
Minutes

Name three consecutive numbers, each less than 100, such that the smallest is divisible by 6, the next is divisible by 5 and the largest is divisible by 4. *(Note: Consecutive numbers are whole numbers that follow in order. An example is 4, 5, 6, 7, 8.)*

✓

54, 55, 56

3E.
6
Minutes

A large cube, 5 cm by 5 cm by 5 cm, is painted orange on all six faces. Then it is cut into 125 small cubes, each 1 cm by 1 cm by 1 cm. How many of the small cubes are *not* painted orange on any face?

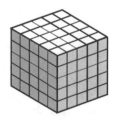

✓

27

4A.
3
Minutes

What is the last year in the twentieth century which is divisible by 37?

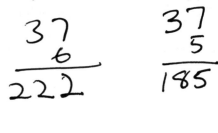

4B.
6
Minutes

In a "Tribonacci Sequence," each number after the third number is the sum of the preceding three numbers. For example, if the first three numbers are 5, 6, and 7, then the fourth number is 18 because $5 + 6 + 7 = 18$ and the fifth number is 31 because $6 + 7 + 18 = 31$. The first five numbers of another Tribonacci Sequence are P, Q, 86, 158, and 291 in that order. What is the value of P?

4C.
5
Minutes

Kim was elected class president. She received 3 votes for every 2 that Amy got. No one else ran. However, if 8 of the people who voted for Kim had voted for Amy instead, Kim would have received only 1 vote for every 2 that Amy would have gotten. How many people voted?

4D.
6
Minutes

Alma has a set of stones. Each weighs a whole number of ounces. By choosing the appropriate stones from the set, she can make any whole number of ounces from 1 to 31 ounces, inclusive. What is the fewest number of stones that Alma can have?

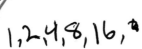

4E.
7
Minutes

Three circular streets intersect at points A, B, C, D, E, and F as shown. How many different paths can be walked along the streets from A to F, if no intersection is entered more than once when walking each path?

5A.
5 Minutes

This morning Allen had 2 more action figures than Barbara, and Barbara had 8 more action figures than Charlie. Allen gave some of his action figures to Charlie, and now both Allen and Charlie have the same number of action figures. Barbara now has N more action figures than Charlie does. What is the value of N?

$$A = B + 2$$
$$B = C + 8$$

5B.
5 Minutes

Frank read nine consecutive pages from a certain book. The sum of the page numbers he read is 378. What is the page number of the middle page he read?

42

5C.
6 Minutes

Tanya forms her sequence by starting with the number 2 and adding 3 continuously. She gets the following: 2, 5, 8, 11, 14, and so on. Joe forms his sequence by starting with the number N (not 2) and adding the number D (not 3) continuously. His second number is 9 and his fifth number is 21. What is his 1000th number?

$$4x - 1 \qquad 9 + 3x = 21 \qquad 5, 9, 13, 17, 21 \qquad \boxed{3999}$$
$$5 + 4(x-1) \quad 4x$$

5D.
5 Minutes

In the multiplication example, different letters represent different non-zero digits and a blank space may represent any digit . What are *both* correct values for the three-digit product DDD?

$$\begin{array}{r} A\,B \\ \times\,C\,D \\ \hline -\,-\,- \\ =\,= \\ \hline D\,D\,D \end{array}$$

$$\begin{array}{r} 33 \\ 123 \\ \hline 3 \end{array}$$

5E.
7 Minutes

Candy and Sandy compete in more than one event. No one else competes. In each event, the winner gets the counting number of points W and the loser gets the smaller counting number of points L. There are no ties. Candy wins the competition by a score of 22 to 13. If Sandy wins exactly one event, what is the value of W?

$$Sw + CL = 13$$
$$Cw + SL = 22$$
$$W = 9$$
$$L = 4$$

SET 4   OLYMPIAD 1

1A.
4 Minutes

For how many different counting numbers between 10 and 200 is the sum of the digits equal to 6, if zero is not a digit of any of the numbers?

9

15, 24, 33, 42, 51,
114, 123, 132, 141, ⚡

1B.
5 Minutes

The perimeter of a rectangle is 26 units. Each side is measured in counting numbers. What is the greatest possible area of the rectangle, in square units?

$2(\ell + w) = 26$
$\ell + w = 13$
40, 42

42 un²

1C.
5 Minutes

Bryan can buy candy canes at 4 for 50¢ and can sell them at 3 for 50¢. How many canes must Bryan sell in order to make a profit of $5.00?

12.5
16.7777

$\frac{40}{4}$ $\frac{4}{50}$ $\frac{3}{50}$

120

1D.
5 Minutes

A supermarket clerk makes a solid pyramid out of identical cereal boxes. The top five layers are shown. What is the total number of cereal boxes in these top five layers?

100

50+50

4
10
18
28
40

1E.
6 Minutes

A normal duck has two legs. A lame duck has one leg. A sitting duck has no legs. Donald has 33 ducks. He has two more normal ducks than lame ducks and two more lame ducks than sitting ducks. How many legs in all do the 33 ducks have?

$n = \ell + 2$
$\ell = s + 2$
$s = s + 4$

Problems - Division E Olympiads

2A.
5 Minutes

Roberta throws five darts at the target shown. Each dart lands in a region of the target, scoring the points shown. Of the following total scores, list all that are *not* possible:
6, 14, 17, 38, 42, 58

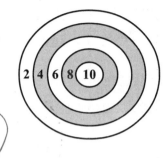

$10 < x < 50$ 6, 17, 58

2B.
4 Minutes

Kelly made two purchases. She gave one cashier $20 for a compact disc and received $6 change. Then she gave another cashier $15 for a bracelet and received $3 change. After these purchases she had $28. How many dollars did she have before buying the compact disc and the bracelet?

63   54   14   12

2C.
5 Minutes

On a 100 cm measuring stick, marks are made at 19, N, and 99 cm, from left to right. The distance between the marks at N and 99 cm is three times the distance between the marks at N and 19 cm. What number is N?

$N = 39$

2D.
5 Minutes

A cube has 6 faces: top, bottom, and all 4 sides. The object shown is made of six congruent cubes. Not all faces are visible. All outer faces of the object including the bottom are painted blue. How many faces of the cubes are painted blue?

$5 + 4 + 4 + 4 + 4 + 5 = 26$

2E.
7 Minutes

Assume that a post office issues only 3¢ and 8¢ stamps and all postage is in whole numbers of cents. What is the greatest amount of postage in cents which *cannot* be made using only 3¢ and 8¢ stamps?

Guess = 96¢

3A.
4 Minutes

Suppose Sandy writes every whole number from 1 to 100 without skipping any numbers. How many times will Sandy write the digit "2"?

(20) 2, 12, 22, 20, 21, 23, 24, 25, 26, 27, 28, 29
32, 42, 52, 62, 72, 82, 92

3B.
4 Minutes

Paul has half as many pieces of candy as Jennifer. Jennifer has half as many as Charles. Charles has 12 times as many as Susan. Susan has 4 pieces. How many pieces do Charles and Paul have altogether?

12 24 48 48
(60) $\frac{48}{2} = \frac{24}{2} = $ (12)

3C.
5 Minutes

Consider all pairs of counting numbers whose sum is less than 11. The two members of a pair could be either the same as each other or different. How many different products are possible if the two numbers are multiplied?

3,3 (3,6)   4,4   (5,5) (1,1) (1,6) 2,2 (2,7)   1 2 3 4 5
3,4 (3,6,7) (4,5)   (1,1) (1,7) 2,3 (2,8)   9 7 5 3 1
(3,5)   (4,6)   (1,3) (1,8) 2,4
                         (1,4) (1,9) (2,5)   (19)
             (1,5)   (2,6)

3D.
6 Minutes

Mary and Kevin each have a rectangular garden whose area is 36 square meters. Each side is measured in whole meters. Mary's garden is 1 m wider than Kevin's garden, but Kevin's garden is 3 m longer than Mary's garden. How wide is Mary's garden, in meters?

$lw = 36$
$(l-3)(w+1) = 36$

$lw = 36$
$lw - 3w + l - 3 = 36$
$lw - 3w + l = 39$

3E.
7 Minutes

Main Street has eight traffic lights. Each shows green for 2 minutes, then switches to another color. The traffic lights turn green 10 seconds apart, from the first light to the eighth light. From the time that the first light turns green until it switches to another color, for how long will all eight lights show green at the same time, in seconds?

4A.
4
Minutes

Each of *PQ* and *RS* represents a 2-digit number. Different letters represent different digits, chosen from 6, 7, 8, and 9. What is the greatest product that *PQ* × *RS* can have?

8342   PQ RS
6,7,8,9

97
86₂

97
86
582
776
8342

4B.
4
Minutes

Exactly 10 disks are in a bowl. Each is marked with a different counting number selected from 1 through 10. Gina and Monique each selects 5 disks. Two of Gina's disks are marked 2 and 8. Two of Monique's disks are marked 7 and 9. What is the largest sum that Gina's disks can have?

4 5 6 2 8 ㉕ _ _ 79

4C.
5
Minutes

A "fast" clock gains time at the same rate every hour. It is set to the correct time at 10 AM. When the fast clock shows 11 AM the same day, the correct time is 10:52 AM. When the fast clock shows 3:30 PM that day, what is the correct time?

12:00 – 11:44
1:00 – 12:36   1 hour – 8 min
2:00 – 1:24
3:00 – 2:16   330 – 2:92

4D.
6
Minutes

All counting numbers are arranged in the triangular pattern as shown by the first four rows. What is the first number in the 13th row?

```
          1
        2  3  4
      5  6  7  8  9
   10 11 12 13 14 15 16
                  ...and so on.
```

⑭⑤ 4

4E.
6
Minutes

A 14-digit number *N* is created by writing 8 as both the first and last digits and then placing the 3-digit number 793 between the two 8s four times. What is the remainder when *N* is divided by 7?

5A.
4 Minutes

In this addition different letters represents different digits. What digits do *A*, *B*, *C*, and *D* represent?

A=3   D=5
B=7   C=8

$$\begin{array}{r} 6\,B\,5\,2 \\ 9\,C\,4 \\ +\,A\,3\,7\,D\,5 \\ \hline 1\,1\,1\,1\,1 \end{array}$$

5B.
4 Minutes

Each of *AB* and *BA* represents a two-digit number having the same digits, but in reverse order. If the difference of the two numbers is 54 and *A* + *B* = 10, find both numbers, *AB* and *BA*.

82, 28

$$\begin{array}{r} 82 \\ 28 \\ \hline 54 \end{array}$$

5C.
5 Minutes

As shown, the length of each segment in the overlapping rectangles is given, in cm. Find the sum of the areas of the shaded regions, in sq cm.

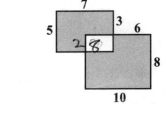

99 cm²

27
72

5D.
6 Minutes

How many different counting numbers less than 200 are exactly divisible by either 6 or 9 or by both?

33   198   22   3×6=18
22   44   11   33×6=198

5E.
6 Minutes

In the number 203,500, the last two zeroes are called *terminal zeroes*. The zero after the digit 2 is not a terminal zero. How many terminal zeroes does the product of the first 30 counting numbers (1 × 2 × 3 × . . . × 30) have?

SET 5 OLYMPIAD 1

1A.
4 Minutes

To engrave a word on a metal plate, each of the first three letters costs 5 cents, and each letter after the first three costs 1 cent more than the preceding letter does. Find the cost, in cents, of engraving the following word:

OLYMPIADS

66¢

15
15
15
21
✓

1B.
5 Minutes

Each of three Math Team students practices at a different time: 10:00 AM, noon, and 2:00 PM. Jean practices either at noon or at 2:00 PM. Krisha doesn't practice at noon, Lisa doesn't practice at 2:00 PM, and Krisha practices 2 hours before Lisa. At what time does each girl practice?

Jean = 2 pm
Krishna = 10 am
Lisa = noon

✓

1C.
5 Minutes

Find the least counting number which when divided by 6 gives a remainder of 1, and when divided by 11 gives a remainder of 6.

17
$y = 11x + 6$
$y = 6x + 1$

$6x + 1 = 11x + 6$
$-5x = 5$
$x = -1$

1D.
5 Minutes

The figure shown consists of 6 congruent squares. The area of the figure is 96 sq cm. What is the perimeter of the figure, in cm?

14
× 4
56

56 cm

✓

1E.
6 Minutes

One seat in an auditorium is broken. It is in the 3rd row from the front of the auditorium and in the 18th row from the back of the auditorium. There are 12 seats to its left and 17 seats to its right. If every row has the same number of seats, what is the total number of seats in that auditorium?

600   20

✓

2A.
3
Minutes

Kristen has had her cat since it was a kitten. She said, "If you multiply my cat's age by 4, and then divide by 12, you get 5. How old is my cat?"

15 years ✓

2B.
4
Minutes

Amy had a nickel, a dime, a quarter, a half-dollar, and a silver dollar. After she lost one coin, she had exactly seven times as much money as her brother had. Which coin did she lose?

5   10   25   50   100

Half-dollar 190 ✓

2C.
5
Minutes

In a four-digit number, the sum of the thousands and hundreds digits is 3. The tens digit is 4 times the hundreds digit. The ones digit is seven more than the thousands digit. No two digits are equal. What is the four-digit number?

2
1

2D.
5
Minutes

A square piece of paper is folded in half to form a rectangle. This rectangle has a perimeter of 24 cm. Find the area of the original square, in sq cm.

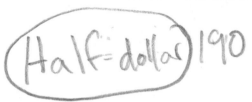

64 cm² 4 + 4

4 ☐ 8

2E.
6
Minutes

Boston is 295 miles from New York City along a certain route. A car starts from Boston at 1:00 PM and travels along this route toward New York at a steady rate of 50 mph. Another car starts from New York at 1:30 PM and travels along this route toward Boston at a steady rate of 40 mph. At what time do the cars pass each other?

295

1:30	25	0
2:00	50	20
2:30	75	40
3:00	100	

7.58 minutes

3A.
4 Minutes

For every two widgets I buy at the regular price, I can buy a third widget for $4. I bought nine widgets for a total of $90. Find, in dollars, the regular price of a widget.

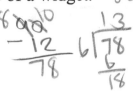

$90 / -12 / 78$ $13 \over 6 \overline{)78}$ ✓ $\boxed{\$13}$

3B.
4 Minutes

On his birthday, Newton was 14 years old and his father was 41. Newton noticed that his age was his father's age with the digits reversed. How many years later will their ages next have their digits reversed?

$-25 \over 14$ $-52 \over 41$ ✓ $\boxed{\text{In 11 years}}$
11 $6 1$

3C.
5 Minutes

A standard clock is set correctly at 1:00 PM. If it loses 3 minutes every hour, what will the clock show when the correct time is 10:00 AM the next day?
(*Note: the number of minutes in your answer must be less than 60.*)

27 $-17 \over 53$ $\boxed{9:33} \text{ A.M}$ X

3D.
5 Minutes

Each side of a square is 6 cm long. A rectangle has a width of 3 cm, and the same area as the square. Find the perimeter of the rectangle, in cm.

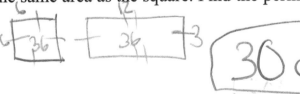

$\boxed{30 \text{ cm}}$ ✓

3E.
6 Minutes

In a set of counting numbers, all have different values. Their sum is 350. Their average is 50. One of the numbers is 100. What is the greatest number that can be in the set?

$350 \div 50$ ⑦ $100 + 245 + 1 + 1 + 1 + 1 +)$ X $\boxed{245}$

4A.
4
Minutes

The 14 digits of a credit card number are written in the boxes below. If the sum of any three consecutive digits is 20, What digit is in Box *A*?

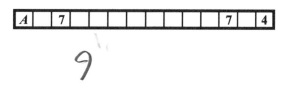

| A | | 7 | | | | | | | | | 7 | | 4 |

9

4B.
5
Minutes

A grocer bought 15 dozen oranges at $1.00 a dozen. She threw away 20 rotten oranges, and then sold the rest at 8 oranges for 85 cents. How much profit did the grocer make, in dollars and cents?

2

4C.
5
Minutes

Mark has 42 identical cubes, each with 1-cm edges. He glues them together to form a rectangular solid. If the perimeter of the base is 18 centimeters, find the height of the rectangular solid, in cm.

3

4D.
6
Minutes

In the multiplication example shown, different letters represent different digits. Find the value of *A* and of *B*.

$$\begin{array}{r} A\,8 \\ \times\ 3\,B \\ \hline 2\,7\,3\,0 \end{array}$$

A=7   B=5

4E.
6
Minutes

A firefighter stood on the middle rung of a ladder, went up 3 rungs, was forced down 5 rungs, and then went up 7 rungs to extinguish the fire. Then the firefighter climbed the remaining 6 rungs to the top of the ladder. How many rungs are there on the entire ladder?

23

5A.
4
Minutes

Twins Joanie and Tony each start with the same number of cents. Joanie buys 1 candy bar and has 70 cents left. Tony buys 3 candy bars at the same price and has 20 cents left. How many cents did Joanie start with?

95.

5B.
4
Minutes

In one complete day, a man binds 100 books and his helper binds one-fourth as many books. If they take turns working complete days, how many days would it take them to bind 500 books?

8

5C.
6
Minutes

What is the greatest possible sum that can result from

BAD + MAD + DAM

if different letters represent different digits, chosen from 1, 3, 8, and 9?

2056?

5D.
6
Minutes

A single story house is to be built on a rectangular lot 70 feet wide by 100 feet deep. The shorter side of the lot is along the street. The house must be set back 30 feet from the street. It also must be 20 feet from the back lot line and 10 feet from each side lot line. What is the greatest area that the house can have, in sq ft?

2 500

5E.
5
Minutes

At Hudson High School, the first class starts at 8:26 AM and the fourth class ends at 11:26 AM. There are 4 minutes between classes and each class is the same length. How many minutes are there in one class period?

42

SET 6   OLYMPIAD 1

1A.
3
Minutes

What is the value of the following?

$$268 + 1375 + 6179 - 168 - 1275 - 6079$$

100 + 100 + 100 = 300

1B.
5
Minutes

Each of 8 boxes contains at least one marble. Each box contains a different number of marbles, except for two boxes which contain the same number of marbles. What is the fewest marbles that the 8 boxes could contain in all?

1.2 3 4 5 6 7
1 2 3 4 5 6 7 8

29

1C.
5
Minutes

Find the whole number which is less than 100, a multiple of 3, a multiple of 5, odd, and such that the sum of its digits is odd.

15

4+5=9

45

1D.
6
Minutes

Takeru has four 1-cent stamps, three 5-cent stamps, and three 25-cent stamps. How many different postage amounts of at least 1 cent can Takeru make?

1E.
6
Minutes

This figure is made up of five congruent squares. The perimeter of the figure is 72 cm. Find the area of the figure, in sq cm.

72 ÷ 12 = 6

36
5
180

180 CM²

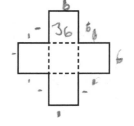

SET 6   OLYMPIAD 2

2A.
4
Minutes

An odd number between 301 and 370 has three different digits. If the sum of its digits is five times the hundreds digit, find the number.

357

357

2B.
4
Minutes

In this subtraction, *PR5T* and *47Y6* represent 4-digit numbers. What number does *PR5T* represent?

6754

2C.
7
Minutes

Ten people stand in line. The first goes to the back of the line and the next person sits down, so that the person who was third is now first in line. Now that person goes to the back of the line and the next person sits down. This process is repeated until only one person remains. What was the original position in line of the only remaining person?

2D.
6
Minutes

Tape, two centimeters wide, is used to completely cover a cube 10 centimeters on each edge. Find the length of tape needed, in centimeters, if there is no overlap of the tape.

12 × 10 = no   120 centimeters

2E.
7
Minutes

The five members of the computer club decided to buy a used computer, dividing up the cost equally. Later, three new members joined the club and agreed to pay their fair share of the purchase price. This resulted in a saving of $15 for each of the original five members. What was the price of the used computer, in dollars?

3A.
3 Minutes

Dr. Bolton was born in an interesting year. The tens digit was twice the thousands digit, the ones digit was three times the tens digit, and the hundreds digit was equal to the sum of the other three digits. In what year was he born?

1 9 2 6
1   hundred ones
thousand tens

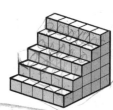
1926

3B.
5 Minutes

Using this staircase of 1-cm cubes, how many more 1-cm cubes will you need to make a cube measuring 5 cm on each side?

20 more cubes

3C.
6 Minutes

Numbers such as 543 or 531 have their digits in *decreasing order* because each digit is less than the digit to its left. The digits in 322 are *not* in decreasing order. How many whole numbers between 100 and 599 have their digits in decreasing order ?

3D.
5 Minutes

One light flashes every 2 minutes and another light flashes every 7 minutes. If both lights flash together at 1 PM, what is the first time *after* 3 PM that both lights flash together?

1:02 1:04   6
1:07 1:14   7

3E.
6 Minutes

Richard has 17 coins with a value of 76 cents. The coins are nickels, dimes, and pennies. He has twice as many pennies as dimes. How many nickels does Richard have?

SET 6 OLYMPIAD 4

4A.
3
Minutes

In the figure at the right, the hand now pointing to 5 moves clockwise to the next numeral every hour. To which numeral will the hand point 24 hours from now?

(1) ✓

4B.
6
Minutes

A square contains two squares, *A* and *B*, with areas of 16 square meters and 9 square meters, as shown. Find the number of meters in the perimeter of the shaded region.

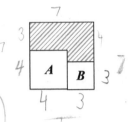

28 meters ✓

4C.
5
Minutes

The average of six numbers is 4. A seventh number is added and the new average is 5. Find the seventh number.

11

4D.
6
Minutes

The numbers 1, 2, 3, and 4 are placed in the empty squares so that each of the four numbers appears in each row, in each column, and in each diagonal. Find the value of *A* and the value of *B*.

1	2	3	4
3	4	1	2
4	3	2	1
A	B	4	3

A = 2   B = 1 ✓

4E.
7
Minutes

Early one morning Sandy took out one-half of the coins from her coin bank, and in the evening she put in 20 coins. The next morning she took out one-third of the coins in the bank, and that evening she put in 4 coins. The next morning, she took out one-half the coins in the bank, leaving 15 coins. How many coins were in the bank to begin with?

Problems - Division E Olympiads

5A.
3 Minutes

Bob collects stamps. Each day he adds 4 stamps to his collection. At the end of three days he has 50 stamps. How many stamps does he have at the end of 10 days?

5B.
5 Minutes

The odometer in Mr. Jackson's car shows he has traveled 62,222 miles. What is the least number of additional miles that Mr. Jackson must travel before the odometer again shows 4 of the 5 digits the same as each other?

5C.
6 Minutes

Exactly one of the numbers 2, 3, 5, and 7 is placed in each of the 4 empty regions formed by circles A, B, and C so that the sum of the numbers in each circle is the same. Each number is used. What is the sum of the numbers in each circle?

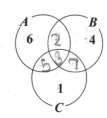

5D.
7 Minutes

Twenty unit cubes are glued together to form this figure, with "holes" which you can see through. The total figure measures $3 \times 3 \times 3$. If the figure is fully dipped in a bucket of paint, how many square units of surface area would be painted?

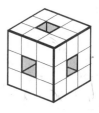

5E.
5 Minutes

The Patriots beat the Giants in a football game. The sum of their scores was 44. The difference of their scores was 20. How many points did the Patriots score?

1A.
3
Minutes

What is the value of the product?

$$25 \times 17 \times 4 \times 20$$

(handwritten work)
17
34
34000
100 × 340
34000

1B.
4
Minutes

4 people can be seated at a single card table. If two tables are placed end to end, 6 people can be seated as shown in the diagram. How many tables must be placed end to end to seat 22 people?

(handwritten) 10 tables ✓

1C.
5
Minutes

A group consisted of 2 girls for every boy. 24 more girls joined the group. There are now 5 girls for every boy. How many boys are in the group?

(handwritten work) 5·16 9   8   b   8 boys   16   40 9   8 b   ✓

1D.
6
Minutes

One owl hoots every 3 hours, a second owl hoots every 8 hours, and a third owl hoots every 12 hours. If they all hoot together at the start, how many times during the next 80 hours will *just two* owls hoot together?

1E.
6
Minutes

Amy, Dawn, and Soumiya sit at a round table facing the center. Each girl wears one ring. Each ring is a different color. One ring is red, one is green, and one is blue. The girl with the green ring is at Amy's right. The girl with the red ring is at Soumiya's left. State the color of each girl's ring.

(handwritten) D - O - A   S = Green Amy = Red D = Blue

SET 7 OLYMPIAD 2

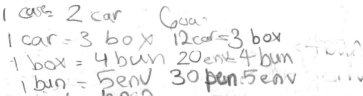

2A.
4
Minutes

One case holds 2 cartons. Each carton holds 3 boxes. Each box holds 4 bundles. Each bundle holds 5 envelopes. Each envelope holds 6 pencils. What is the greatest number of pencils that one case can hold?

1 case 2 car   6 var
1 car = 3 box   12car=3 box
1 box = 4 bun   20 env 4 bun
1 bun = 5 env   30 pen 5 env
1 env = 6 pen

2B.
5
Minutes

A factory has enough oil on hand to last 30 days if 4 barrels of oil are used each day. How many barrels should be used each day if the same amount of oil is to last 40 days?

30 : 4   15 : 2

2C.
5
Minutes

Find the 4-digit number *ABCD* if:

$$
\begin{array}{r}
A\,B\,C\,D \\
\times \qquad 9 \\
\hline
D\,C\,B\,A
\end{array}
$$

2D.
6
Minutes

How many multiples of 7 are there between 100 and 1000?

2E.
6
Minutes

A garden, 10 m by 20 m, is enclosed by a sidewalk of width 1 meter. Find the area of the sidewalk, in sq m.

4/5

3A.
4
Minutes

Find a two-digit number with all these properties:
- the tens digit is larger than the ones digit,
- the difference between the digits is greater than 3,
- the sum of the digits is greater than 10, and
- the number is a multiple of 12.

12, 24, 36, 48, 60, 72, 84, 96

84

2/5

3B.
5
Minutes

A shopkeeper sells house numbers. She has a large supply of the numerals 4, 7, and 8, but no other numerals. How many different three-digit house numbers could be made using only the numerals in her supply?

478 748 847
487 784 874

6

3C.
5
Minutes

Find the value of the following.

$$100 - 98 + 96 - 94 + 92 - 90 + \ldots + 8 - 6 + 4 - 2$$

2 + 2   12

3D.
5
Minutes

The interior of a circle can be cut into a maximum of 2 regions by one straight line. The interior can be cut into a maximum of 4 regions by 2 straight lines. What is the maximum number of regions into which the interior can be cut by 3 straight lines?

7

3E.
6
Minutes

A theater has 18 rows of seats. Each row has the same number of seats. Suppose there were 3 fewer rows. Then for the theater to hold the same number of people, each remaining row would need 6 more seats. How many seats are in the theater?

36

4A.
5
Minutes

Amy has three older sisters. Beth is 2 years older than Amy. Jo is 3 years older than Beth. Meg is 4 years older than Jo. Meg is twice as old as Beth. How old is Amy?

4B.
5
Minutes

Suppose that *A*, *B*, and *C* represent 1, 2, and 3 in some order. What is the greatest possible sum that can result from this addition?

```
  A B C 4
5 A B C 0
+ C B 6 A 8
```

4C.
5
Minutes

A gumball machine contains red, green, yellow, and purple gumballs. You cannot control which color you get. Anna wants three gumballs of any one color. At 5¢ each, what is the minimum number of cents that guarantees in advance three gumballs of the same color?

4D.
5
Minutes

The area of a rectangle is 36 sq cm. The lengths of the sides are whole numbers of centimeters. What is the greatest perimeter the rectangle can have, in cm?

4E.
6
Minutes

The I. M. Hipp Tire Company produces tires for cars and two-wheel motorcycles. One week the company produced a total of 269 tires for 70 vehicles. This included a spare tire for each car but not for any motorcycle. How many motorcycle tires did the company produce that week?

5A.
3 Minutes

Toni picks a number and multiplies it by 3. She then adds 4 to the result and finally divides this new number by 2. Her final result is 14. With what number did she start?

5B.
4 Minutes

Suppose five days before the day after tomorrow was Friday. What day was yesterday?

5C.
6 Minutes

If 3 smiles = 10 grins and 6 grins = 9 laughs, how many laughs does it take to equal 2 smiles?

5D.
6 Minutes

Anna wants to cover the outside of a rectangular box with colored paper. The box has a square base with area of 16 square inches. The volume of the box is 80 cubic inches. How many square inches of paper will Anna need to completely cover the box, including the top and bottom, with no paper left over?

5E.
6 Minutes

The digits 1, 2, 3, 4, and 5 are each used once to write a five-digit number *ABCDE*. The three-digit number *ABC* is divisible by 4, *BCD* is divisible by 5, and *CDE* is divisible by 3. Find the five-digit number *ABCDE*.

SET 8   OLYMPIAD 1

4
5

4
5

1A.
4
Minutes

What number does N represent?

$42 \times 25 = 21 \times N$

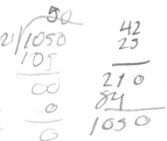

√ $N: 50$

1B.
5
Minutes

Find the number between 100 and 999 such that
- the sum of the three digits is 12.
- the hundreds digit equals the sum of the tens digit and the ones digit.
- the hundreds digit is one greater than the tens digit.

651   √   651

1C.
5
Minutes

For every \$3 Marisa spends, Andie spends \$5. Andie spends \$120 more than Marisa does. How many dollars does Andie spend?

1D.
4
Minutes

A rectangle 4 cm by 12 cm is divided into four triangles, as shown. The areas of three of the four triangles are stated in the diagram. Find the number of sq cm in the area of the shaded triangle.

√   6 sq. cm

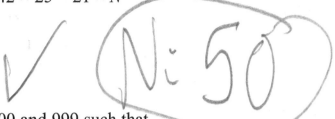

2
18
16
-8
42

1E.
6
Minutes

Jeff has some coins. He can put the same number of coins into each of 9 bags, with no coins left over. He can put the same number of coins into each of 6 bags, with no coins left over. He can put the same number of coins into each of 5 bags, with 2 coins left over. What is the least number of coins Jeff could have?

72 coins

SET 8   OLYMPIAD 2

2A.
4
Minutes

The sum of five consecutive whole numbers is 45. What is the least of the five numbers?

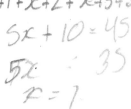

$x + x+1 + x+2 + x+3 + x+4 = 45$

$5x + 10 = 45$

$5x = 35$

$x = 7$

⑦

2B.
5
Minutes

Sara places four books on a shelf. The blue book must be somewhere to the left of the green book. The red book must be somewhere to the left of the yellow book. In how many different orders can Sara place the books?

r b g y   b r y g
r y b g   b r g y
r b y g   b g r y

⑥

2C.
5
Minutes

Ann, Bill and Chris each take two Vitamin C tablets daily. Wendy takes only one tablet daily. One bottle has enough tablets to last these four people exactly 24 days. Suppose Wendy takes two tablets daily, instead of one tablet. How many days will the tablets in one bottle last?

2D.
5
Minutes

The perimeter of a rectangle is 14 m. All its side-lengths are whole numbers of meters. List ALL possible numbers of square meters which could be the area of the rectangle.

6, 10, 12,

1,1,6,6
2,2,5,5
3,3,4,4
0,4,3,3

2E.
5
Minutes

Aiah lists all the counting numbers from 1 through 200. How many times will the digit 4 appear on Aiah's list?

1-10   11-20   20-30   30-40   41-50   51-60
1   1   1   2   10   1

20 + 20 = ⟨ 40 times ⟩

SET 8   OLYMPIAD 3

3A.
4 Minutes

A starting number is multiplied by 4. Then 14 is added to the result. The new number is 6 times the starting number. What is the starting number?

7

3B.
5 Minutes

The first time Hannah plays a video game, she loses in 30 seconds. Hannah improves so that each game she plays is 30 seconds longer than the game before. Hannah's final game is 4 minutes long. It costs her 25¢ to play each game. How much does Hannah spend?

2

3C.
6 Minutes

Dwayne's street has 12 houses on it. Every day more letters are delivered to his house than to any of the others. Today 57 letters were delivered to his street. What is the minimum number of letters that Dwayne could have received today?

6

3D.
5 Minutes

A rectangle is divided up into four smaller rectangles. The length of each segment is a whole number of cm. The areas of three of the rectangles are given in the diagram. Find the number of sq cm in the total area of rectangle *ACEG*.

A		C
6 sq cm	10 sq cm	
9 sq cm		
G		E

40

3E.
6 Minutes

Three darts are thrown at the dartboard shown. A miss scores 0 points. The three scores are added together. Find the least whole number total score that is *impossible* to obtain.

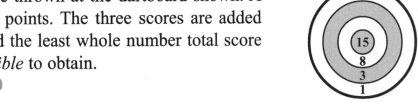

13?

SET 8   OLYMPIAD 4

4A.
4 Minutes

Madison buys some five-cent stamps and some ten-cent stamps for a total of $1.00. She receives 13 stamps. How many five-cent stamps does she buy?

6

4B.
5 Minutes

How many even 2-digit numbers have an *odd* number as the sum of their digits?

25

4C.
6 Minutes

If I start with 3 and count by 4's until I reach 99, I get 3, 7, 11, …, 99, where 3 is the first number, 7 is the second number, 11 is the third number, and so on. If 99 is the Nth number, what is the value of N?

25

4D.
5 Minutes

Thirty cubes are placed in a line such that they are joined face to face. The edges of each cube are one cm long. Find the surface area of the resulting solid, in sq cm.

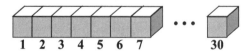
1 2 3 4 5 6 7   30

4E.
5 Minutes

How many different counting numbers will each leave a remainder of 5 when divided into 47?

5

5A.
4
Minutes

Kristen puts boots on her four cats. Each front leg has a boot with 4 eyelets. Each rear leg has a boot with 6 eyelets. Each of the four cats loses a different boot. What is the total number of eyelets in the remaining boots? (*Note: An eyelet is a small hole for a shoelace*)

60

5B.
5
Minutes

Two years ago my age was a multiple of 6. Last year it was a multiple of 5. I am less than 50 years old. How old am I now?

26

5C.
6
Minutes

The five points shown represent five towns. Kevin starts at town *A* and visits each of the other towns exactly once. In how many different orders can Kevin visit the other four towns?

24

5D.
7
Minutes

An empty carton is opened and flattened to form the figure shown. The carton has both a top and a bottom. Find the total area of the figure shown, in sq cm.

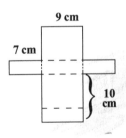

222

5E.
5
Minutes

5*A*4 and 6*B*9 are three-digit numbers. 6*B*9 is the sum of 5*A*4 and 125. 6*B*9 is divisible by 9. Find the digit *A*.

1

3/5

SET 9   OLYMPIAD 1

1A.
3
Minutes

Kim stands in a line of people. She is the 25th person counting from the front of the line. She is the 12th person counting from the rear. How many people are in the line?

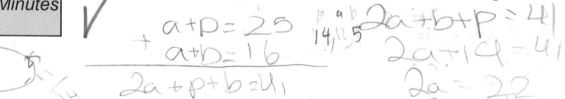

36 people

$$\begin{array}{r} 25 \\ 12 \\ \hline 37 \end{array}$$

✓

1B.
5
Minutes

There are 5 girls in a tennis class. How many different doubles teams of 2 girls each can be formed from the students in the class?

10

1C.
5
Minutes

At a fruit stand, an apple and a pear cost 25 cents, a pear and a banana cost 19 cents, and an apple and a banana cost 16 cents. Alex buys one apple, one pear, and one banana. How much does Alex spend, in cents?

30¢

$$\begin{array}{r} a+p = 25 \\ + \quad a+b = 16 \\ \hline 2a+p+b = 41 \end{array}$$

$2a+b+p = 41$
$2a+19 = 41$
$2a = 22$

$\begin{array}{r} 41 \\ -19 \\ \hline 22 \end{array}$

$\frac{22}{22}$

1D.
6
Minutes

A rectangle has a perimeter of 90 cm. The length of the rectangle is 25 cm more than its width. Find the area of the rectangle, in sq cm.

340

$2w + 2w + 50 = 90$
$4w + 50 = a0$
$4w = 40$
$w = 10$

35 cm²

1E.
6
Minutes

Find the sum of the *digits* of the first 25 odd counting numbers.

1, 3, 5, 7, 9, 11, 13, 15, 17, 19 → 65
21, 23, 25, 27, 29, 31, 33, 35, 37, 39 → 145
41, 43, 45, 47

95
145
340

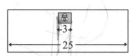

2A.
3 Minutes

A picture 3 feet across hangs in the center of a wall that is 25 feet long. How many feet from the left end of the wall is the left edge of the picture?

2B.
4 Minutes

David buys a Beanie Baby. He later sells it to Jessica and loses $3 on the deal. Jessica makes a profit of $6 by selling it to Bryan for $25. How much did David pay for the Beanie Baby?

2C.
5 Minutes

Ashley is twice as old as Carlos. Billy is 5 years younger than Ashley. The sum of the ages of the three children is 25. How old is Carlos, in years?

2D.
6 Minutes

The whole numbers from 100 down to 0 are arranged in columns P, Q, R, S, and T as indicated. Write the letter of the column that contains the number 25.

P	Q	R	S	T
	100	99	98	97
93	94	95	96	
	92	91	90	89
85	86	87	88	
	84	83	82	81
77	78	79	80	
			and so on ...	

2E.
6 Minutes

A cubical box without a top is 4 cm on each edge. It contains 64 identical 1 cm cubes that exactly fill the box. How many of these small cubes actually touch the box?

SET 9   OLYMPIAD 3

3A.
4
Minutes

T-shirts cost $8 each. Amy buys 1 T-shirt, Becky buys 3, Colin buys 5, Dan buys 7, and Emily buys as many shirts as Amy, Becky, and Colin combined. How many dollars do the five spend in total?

$256

3B.
5
Minutes

Find the sum of all counting numbers less than 25 which are *not* divisible by 2 or 5.

1, 3, 7, 9, 11, 13, 17, 19, 21, 24

124

3C.
5
Minutes

In a group of 25 girls, 8 joined the track team, 13 joined the math team, and 6 joined both teams. How many of the girls did not join either team?

10 girls

3D.
6
Minutes

I climb half the steps in a staircase. Next I climb one-third of the remaining steps. Then I climb one-eighth of the rest and stop to catch my breath. What is the least possible number of steps in the staircase?

24

3E.
6
Minutes

ABCD and *EBCF* are both rectangles. The length of $\overline{CD}$ is 15 cm. The length of $\overline{BC}$ is 8 cm. The length of $\overline{AE}$ is 12 cm. Find the total number of sq cm in the areas of the shaded regions.

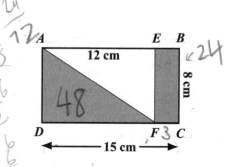

42 cm²

Problems - Division E Olympiads

SET 9 OLYMPIAD 4

4A.
3
Minutes

Each of 6 piles contains a different number of pennies. Each pile contains at least one penny. What is the least possible total number of pennies in the 6 piles?

1,2,3,4,5,6

21

4B.
4
Minutes

The ages of Amanda, Brittany, and Carly are each prime numbers. Amanda is the youngest. The sum of the ages of Amanda and Brittany is equal to Carly's age. How old is Amanda?

2, 3, 5

A B C

2

4C.
5
Minutes

A bus can hold 24 adults or 30 children. 25 children are already on the bus. What is the greatest number of adults that can still get on the bus?

20

4D.
6
Minutes

ABCD is a square. *DEFG* is a rectangle. The length of $\overline{EF}$ is 25 cm. *D* is the midpoint of $\overline{CE}$. The perimeter of the entire shaded region is 180 cm. Find the length of $\overline{AG}$, in cm.

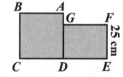

4E.
7
Minutes

The average of a group of 20 numbers is 20. The average of a different group of 60 numbers is 60. The two groups of numbers are combined into a single group. What is the average of the combined group?

20 two's   400
60 60's   3600
          4000

4000÷80 = 50   50

SET 9 OLYMPIAD 5

5A.
4
Minutes

May 25, 2025, will occur on a Sunday. On which day of the week will May 1, 2025 occur?

[handwritten: 1 2 3 4 5 6 7 8 9 10 11 12 13 14 15 16 17 — Th F Sat Su M T W Th F Sa Su M T W Th F Sa — 18 19 20 21 22 23 24 25 — M T W Th Fr Sa Su]

[handwritten answer: Thursday]

5B.
5
Minutes

The four-digit number *A*7*A*8 is divisible by 9. What digit does *A* represent?

[handwritten: A7A8, 15, 6]

5C.
7
Minutes

In each turn of a certain game, only the following point-scores are possible: 5, 3, 2, 0. Eight turns are taken. In how many ways can the total point-score be 25?

(Do not consider the order in which points are scored.)

5D.
7
Minutes

What is the largest number of these wooden "Els" that can be packed in a box that is 2 cm by 4 cm by 6 cm?

[handwritten: 48]

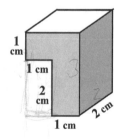

5E.
5
Minutes

Kayla has *N* marbles. She groups them by threes and has one left over. She groups them by sevens and has four left over. Kayla has more than five marbles. What is the least number of marbles Kayla could have?

[handwritten answer: 25]

SET 10   OLYMPIAD 1

1A.
4
Minutes

What is the value of $(185 + 278 + 579) - (85 + 178 + 279)$?

$$500$$ ✓

$185 + 278 + 579 - 85 - 178 - 279$

$100 + 100 + 300$

1B.
4
Minutes

A piece of paper is cut out along the solid lines and labeled as shown. It is folded along the dotted lines to make an open box. The box is placed so that the top is open. What letter is on the bottom of the box?

$\boxed{W}$ ✓

1C.
4
Minutes

Alexis, Emma, and Li play in the school band. One plays the flute. One plays the saxophone. One plays the drums. Alexis is a 6th grader. Alexis and the saxophone player practice together after school. Emma and the flute player are 5th graders. Who plays the drums?

A - ✗DF
E - DS✗
L - FS✗

Alexis ✓

1D.
5
Minutes

Avi uses 11 toothpicks to form a row of 5 attached triangles, as shown. Suppose he continues this pattern, using 89 toothpicks in all. What is the total number of triangles formed?

1E.
7
Minutes

Dale travels from city A to city B to city C and back to city A. Each city is 120 miles from the other two. Her average rate from city A to city B is 60 mph. Her average rate from city B to city C is 40 mph. Her average rate from city C to city A is 24 mph. What is Dale's average rate for the entire trip, in miles per hour?

2A.
4
Minutes

An even number between 100 and 125 is divisible by 3 and also by 5. What is that number?

2B.
5
Minutes

Dom has $120 in his bank account. He deposits $6 at the end of each week. Kim has $200 in her account. She withdraws $4 at the end of each week. At the end of how many weeks will they have the same amount in their accounts?

7 weeks

2C.
6
Minutes

Amanda, Beth, and Sarah run three races. In each race, one of them earns 5 points, one of them earns 3 points, and one of them earns 1 point. After the three races Beth has the highest point total. What is the *least* total score that Beth can have?

2D.
5
Minutes

All three horizontal rows of this grid have the same sum after just two numbers are exchanged. What are the two numbers that are exchanged?

16	8	22
23	4	12
11	27	15

11
23
15
49

2E.
6
Minutes

An ant sits at vertex *V* of a cube with edge of length 1 m. The ant moves along the edges of the cube and comes back to vertex *V* without visiting any other point twice. Find the number of meters in the length of the longest such path.

8 meters

3A.
4 Minutes

Points *A*, *B*, *C* and *D* are on a line. They are not necessarily in that order. Point *A* is between *B* and *C*. Point *B* is between *A* and *D*. Point *D* is to the left of point *C*. List the points in order from left to right.

3B.
5 Minutes

Each day Jeffrey earns $3 for doing certain chores. He can earn $5 instead by also doing additional chores. After ten days of doing chores, Jeffrey has earned a total of $36. On how many of these days did Jeffrey do additional chores?

3 days

3C.
5 Minutes

A four-digit number is written on a piece of paper. Ashley spills whiteout on it. Now the last two digits are no longer visible:

8 6 ? ?

The four-digit number is divisible by three, by four, and by five. Find the four-digit number.

8670

3D.
6 Minutes

In the diagram, angles *A* and *D* are right angles. *AB* = 4 cm, *AD* = 6 cm, and *CD* = 8 cm. Find the area of *ABCD* in square centimeters.

36 cm ✓ Answer

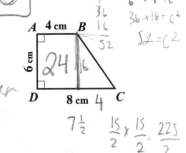

3E.
7 Minutes

Two watches are set correctly at 7:00 AM. One watch gains 3 minutes every two hours. The other watch loses 1 minute every two hours. At what time the next day will the faster watch be exactly one hour ahead of the slower watch? (Indicate AM or PM.)

SET 10   OLYMPIAD 4

4A.
4 Minutes

What is the least counting number that can be added to 259 so that the result is a multiple of 25 ?

16

4B.
5 Minutes

A fair die is thrown. What is the probability that the top face shows a factor of 6?

1/6

4C.
5 Minutes

In this subtraction the squares contain the digits 3, 4, 5, and 8 in some order and the circles contain the digits 1, 3, 7, and 9 in some order. What four-digit number is represented by the squares?

$$\begin{array}{cccc} \square & \square & \square & \square \\ - \bigcirc & \bigcirc & \bigcirc & \bigcirc \\ \hline 3 & 6 & 9 & 9 \end{array}$$

5438

4D.
6 Minutes

Each side of the 9 cm by 9 cm square shown is divided into three equal parts. Find the area of the shaded region, in sq cm.

← 9 →

63

4E.
7 Minutes

Colored beads are placed in the following order: 1 red, 1 green; then 2 red, 2 green; then 3 red, 3 green; and so on. In all, how many of the first 100 beads are red?

55

5A.
5
Minutes

What number multiplied by itself is equal to the product of 6 and 150?

30

5B.
5
Minutes

Jessie has $5.10 worth of stamps. She has equal numbers of 50-cent, 20-cent, 10-cent, and 5-cent stamps. She has no other stamps. How many 50-cent stamps does she have?

6

5C.
6
Minutes

What is the perimeter in meters of the figure shown? All angles are right angles.

42 m

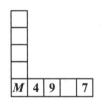

5D.
7
Minutes

The numbers 1 through 9 are placed one per square in the figure. The total of the 5 numbers in the horizontal row is the same as the total of the 5 numbers in the vertical column. Find all the different values that *M* can be.

1, 3

| M | 4 | 9 | | 7 |

5E.
5
Minutes

A bag contains some marbles, all of the same size. Eight of them are black. The rest are red. The probability of drawing a red marble from the bag is $\frac{2}{3}$. Find the total number of red marbles in the bag.

16

OLYMPIAD PROBLEMS

DIVISION M SETS 11-17

SET 11   OLYMPIAD 1

1A.
4 Minutes

The figure at the right is folded on the thin lines to form a cube. What is the *greatest sum* that can be found by adding the three numbers on the three faces which meet at a corner?

(14)

1B.
5 Minutes

What is the mean (average) of all the counting numbers less than 100 which are multiples of 3?

99

1C.
5 Minutes

Each of three darts lands in a numbered region of the dart board, scoring the number of points shown. How many different sums are possible for the three darts?

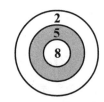

(8, 5, 2)

(8, 8, 2)

8,2,2

1D.
5 Minutes

The product of 45 and the counting number N is a perfect cube. Find the least possible value for N.
(*Definition: a perfect cube is the product of three identical factors. For example: 64 is a perfect cube because $64 = 4 \times 4 \times 4$.*)

1E.
6 Minutes

Two identical squares with sides of length 10 centimeters overlap to form a shaded region as shown. A corner of one square lies at the intersection of the diagonals of the other square. Find the area of the shaded region, in sq cm.

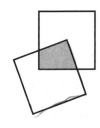

2A.
4 Minutes

Seven congruent rectangles are arranged as shown to form rectangle $ABCD$. If $AB = 20$, find the perimeter of rectangle $ABCD$.

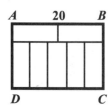

2B.
5 Minutes

The letters in MATH and OLYMPIADS are cycled separately as shown and placed in a numbered list. The next correct spelling of both words MATH and OLYMPIADS will appear in row number n. Find the value of n.

1.	MATH	OLYMPIADS
2.	ATHM	LYMPIADSO
3.	THMA	YMPIADSOL
⋮		⋮ ⋮
n.	MATH	OLYMPIADS

36

2C.
6 Minutes

Suppose P and Q both represent prime numbers such that
$$5 \times P + 7 \times Q = 109$$
Find the value of the prime P.

(*Definition: a prime number is a counting number with **exactly** two different factors: the number itself and the number 1.*)

2D.
6 Minutes

A two-digit number is divided by the sum of its digits. What is the greatest remainder obtained?

2E.
6 Minutes

$ABCD$ is a rectangle as shown. E lies on side $\overline{AD}$ and $\angle BEC$ is a right angle of triangle BEC. $CE = 6$, $BE = 8$, and $BC = 10$. Find the area of rectangle $ABCD$.

SET 11 OLYMPIAD 3

3A.
5 Minutes

Find the *least* whole number N greater than 40 for which:
  (1) N divided by 4 leaves a remainder of 1, and
  (2) N divided by 5 leaves a remainder of 3.

3B.
6 Minutes

Josephine purchased some 25-cent stamps and some 16-cent stamps for a total of $3.62. What is the least number of 16-cent stamps Josephine purchased?

3C.
5 Minutes

The Euclid City School has 1600 students. Each student takes five classes per day, and each teacher teaches four classes per day. Each class contains one teacher and 25 students. How many teachers does the Euclid City School employ?

3D.
6 Minutes

The pages of a book are numbered consecutively from 1 to 100. What is the sum of all the digits used in numbering the pages of the entire book?
(*Note: each digit such as 3 is to be added every time it appears in the numbering.*)

3E.
6 Minutes

Express the extended fraction shown at the right as a simple fraction in lowest terms.

$$\cfrac{3}{4+\cfrac{3}{4+\cfrac{3}{4}}}$$

4A.
5 Minutes

Jakob has 3 more brothers than sisters. How many more brothers than sisters does his sister Sari have?

5

4B.
5 Minutes

Ryan made a deal with Noelle so that he could borrow her calculator. Each time Ryan borrowed Noelle's calculator, Noelle would first triple Ryan's money, and then Ryan would give Noelle twenty-seven cents. Ryan was broke after borrowing Noelle's calculator three times. How many cents did Ryan have at the start?

13

4C.
4 Minutes

There are 25 students on the Geoville Math Team. 11 play chess, 15 play tennis, while 3 play neither chess nor tennis. How many students play chess, but not tennis?

7

4D.
5 Minutes

How many whole numbers between 1 and 150 have *exactly* three different factors? (*Note: For example, 10 has 4 factors: 1, 2, 5, and 10.*)

5

4E.
6 Minutes

A circle rolls once, without slipping, along the outside of a square with sides of length 4 inches, and returns to its starting point at A. The radius of the circle is 1 inch. To the nearest hundredth of an inch, how far does the center of the circle travel? (Use 3.14 for π)

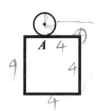

16

SET 11   OLYMPIAD 5

5A.
6
Minutes

Points A, B, C, D, and E are points of the circle as shown. How many different triangles can be formed by joining any three of these points?

(*Note: the triangle formed by joining A to B to C is the same triangle as that formed by joining B to A to C.*)

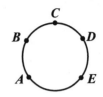

10

5B.
6
Minutes

There are six candidates (A, B, C, D, E, and F) in a school election. The following facts are known to be true: (1) A is elected with exactly 50 votes, (2) B is second, (3) F is last with 5 votes (4) there were a total of 100 votes cast, and (5) no two candidates received the same number of votes. What is the smallest number of votes B could have received?

13

5C.
6
Minutes

If A and B represent digits, then the five-digit number $23AB3$ is exactly divisible by 99. Find the two digit number AB.

A=4 B=6

5D.
5
Minutes

4 chefs require 10 minutes to prepare 20 desserts. At this rate, how many chefs are needed to prepare 75 desserts in 15 minutes?

10

5E.
6
Minutes

Biff's pet is tied to the corner of a square shed, 6 meters on each side. The rope is 8 meters long. The area outside the shed over which the pet can wander is $N\pi$ square meters. Find the whole number N.

SET 12   OLYMPIAD 1

1A.
4
Minutes

Amanda arranges the digits 1, 3, 5, and 7 to write a four-digit number. The 7 is next to the 1, but not to the 5. The 3 is next to the 7, but not to the 5. The four-digit number is divisible by 5. What is Amanda's four-digit number?

3715

1B.
5
Minutes

In this diagram, the big square is divided into 7 congruent small squares and 2 congruent triangles. The area of the shaded square is 4 sq cm. What is the area of the shaded triangle, in sq cm?

18

1C.
5
Minutes

Suppose that $a ❂ b$ means $a + a - b$. For example, $3 ❂ 4$ means $3 + 3 - 4$, which is another name for 2. If $4 ❂ 5$ and $6 ❂ \Box$ represent the same number, what is the value of $\Box$?

1D.
5
Minutes

The average of all five of Carlos' test grades is exactly 84. The average of his first three test grades is exactly 80. What is the average of his last two test grades?

90

1E.
6
Minutes

For 8 weeks of work, Melanie will receive $600 and a new computer. After only 6 weeks of work, she would be entitled to the new computer but only an additional $150. What is the value of the computer, in dollars?

1200

2A.
3 Minutes

If the product of the month and day equals the last two digits of the year, we can call the date an "interesting date". For example, April 9, 1936 (4/9/36) is an interesting date because $4 \times 9 = 36$. For the calendar year 1996, list all interesting dates.

2B.
4 Minutes

In this multiplication example, different letters represent different digits. What is the value of the digit B?

$$\begin{array}{r} 1\,A\,B\,C\,D\,E \\ \times \qquad\qquad 3 \\ \hline A\,B\,C\,D\,E\,1 \end{array}$$

2C.
5 Minutes

As shown, a square is made up of four congruent rectangles and a smaller square. One side-length of the larger square is 14 cm. One side-length of the smaller square is 8 cm. What is the area of the shaded rectangle, in sq cm?

2D.
6 Minutes

A movie ticket costs $4 for children, $6 for senior citizens, and $10 for all others. If 7 people buy tickets, which of the following total sales figures is possible?

$26, $37, $48, $57, $68, $75

2E.
6 Minutes

The symbol 3! means $3 \times 2 \times 1$, which equals 6. Similarly, 5! means $5 \times 4 \times 3 \times 2 \times 1$, which equals 120, and so on. Suppose that $N!$ ends in exactly 3 zeros after fully multiplying out. What is the smallest value that N can have?

3A.
3
Minutes

The three angles of triangle *ABC* (shown at right) are all acute angles. Without adding any new line segments, what is the largest number of acute angles that can be found in this diagram? *(Note: An* <u>acute angle</u> *is any angle whose measure is between 0° and 90°.)*

10

3B.
4
Minutes

List all two-digit numbers that satisfy both of the following:
1. The tens and ones digits are consecutive numbers, and
2. The number itself is the product of two consecutive numbers.

1. 12   2. 56

3C.
4
Minutes

What is the value of *N*?

$$\frac{1}{5} = \frac{1}{6} + \frac{1}{N}$$

30

3D.
5
Minutes

The clock shown at the right loses twelve minutes every hour. It shows the correct time now. In how many hours will it next show the correct time?

60

3E.
7
Minutes

In this diagram, each "path" spells the word OLYMPIAD correctly. How many different paths exist in the diagram?

48

```
        O
      L   L
      Y   Y
        M
      P   P
    I   I   I
      A   A
        D
```

SET 12   OLYMPIAD 4

4A.
3
Minutes

Carol chooses a number. She multiplies it by 4, then adds 8, then divides by 4, and finally subtracts 8. Her end result is 4. What number did she choose?

45

4B.
4
Minutes

What is the greatest odd factor of 4,664?

4C.
5
Minutes

Suppose all counting numbers are arranged in columns as shown at the right. Under what letter will the number 100 appear?

70

A	B	C	D	E	F
		1	2	3	4
10	9	8	7	6	5
11	12	13	14	15	16
22	21	20	19	18	17
23	24	25	26	27	28

... *and so on*

4D.
5
Minutes

Both ABC and $3D8$ are three-digit numbers such that $ABC - 3D8 = 269$. If $3D8$ is divisible by 9, what number does ABC represent?

18

4E.
6
Minutes

Any segment that joins two points of a circle is called a chord of that circle. For example, $\overline{PQ}$ is a chord of the circle shown. Suppose six different chords are drawn in a circle. What is the greatest number of points in which the chords can intersect with each other?

5A.
4 Minutes

How many two-digit counting numbers exist such that the tens digit is larger than the ones digit?

5B.
4 Minutes

A rectangle is divided into four smaller rectangles whose areas in sq cm are 35, 42, 10, and N, as shown. The length of each side of every rectangle is a whole number. What is the value of N, in sq cm?

35	10
42	N

5C.
6 Minutes

A hat contains 20 slips numbered consecutively from 71 to 90. Karen draws 10 slips from the hat including the "74". Rob draws the other 10 slips including the "89". Each adds the numbers written on his or her slips. What is the largest amount by which Karen's total can be greater than Rob's total?

5D.
7 Minutes

All members of the Math Club paid the same amount for their End-of-Year Party. The girls paid a total of $90 and the boys paid a total of $60. However, buying gifts raised the total cost to $210, so each member paid an additional $2. How many members are girls?

5E.
6 Minutes

In the multiplication example $AB \times C = AAA$, different letters represent different digits, AB represents a two-digit number, and AAA represents a three-digit number. What is the value of C?

SET 13 OLYMPIAD 1

1A.
4 Minutes

Suppose it is now the month of November. What month will it be 100 calendar months from now?

1B.
5 Minutes

Ari, Barry, Cara, Dara, and Erin have a total of $85. Ari and Barry have a total of $40. Cara and Dara have a total of $30. Erin and Ari have a total of $20. How many dollars does Barry have?

1C.
5 Minutes

Of all the mathletes at Wantagh Middle School, 80% own computers and 40% are in band. However, 10% of all the mathletes neither own computers nor are in band. What percent of all the mathletes both own computers and are in band?

1D.
5 Minutes

In lowest terms, what is the product of this multiplication?

$$\frac{3}{5}\times\frac{5}{7}\times\frac{7}{9}\times\frac{9}{11}\times\frac{11}{13}\times\frac{13}{15}\times\frac{15}{17}\times\frac{17}{19}\times\frac{19}{21}\times\frac{21}{23}\times\frac{23}{25}\times\frac{25}{27}$$

1E.
6 Minutes

Using only the segments shown at the right, how many triangles of all sizes can be traced?

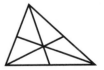

2A.
4
Minutes

A jar has exactly 10 blue marbles, 10 gray marbles, 10 red marbles, and 10 green marbles. If I cannot see the marbles, what is the fewest number of marbles I must pick without replacement in order to *guarantee* that three of the marbles I pick are red?

2B.
5
Minutes

Last week Jessica beat Emily at checkers in 20% of their matches. Emily won 12 matches. How many matches did Jessica win?

2C.
6
Minutes

The five-digit number 839A2 is divisible by 12. What digit does A represent?

2D.
5
Minutes

In simplest terms, what is the value of the following?

$$\frac{6}{.3} + \frac{.3}{.06}$$

2E.
6
Minutes

The large square shown contains smaller squares A, B, and C. Their areas are 9, 16, and 9 sq m, respectively. How many sq m are in the area of the shaded region?

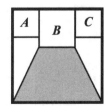

3A.
4 Minutes

Express as a single integer:

$$1 - 3 + 5 - 7 + 9 - 11 + 13 - 15 + 17 - 19$$

3B.
6 Minutes

The cafeteria sells each apple at one price and each banana at another price. For 5 apples and 3 bananas Dan pays $5.70. For 3 apples and 5 bananas Chris pays $4.70. The price of one apple is how much more than the price of one banana, in cents?

3C.
5 Minutes

In this correct multiplication example, different letters represent different digits. What three-digit number is represented by *SUM*?

$$
\begin{array}{r}
U\ M \\
\times\ U\ M \\
\hline
G\ U\ M \\
M\ E \\
\hline
S\ U\ M \\
\end{array}
$$

3D.
5 Minutes

During a trip, Samantha passed mile marker 142 on the highway at 9:10 AM and mile marker 152 at 9:25 AM. Her speed was constant for the whole trip. At what time AM did she pass mile marker 166?

3E.
6 Minutes

A path is traced from point *A* to point *B* by using the segments shown in the diagram at the right. Paths are traced only upwards, to the right, or diagonally upwards. How many different paths can be traced from *A* to *B*?

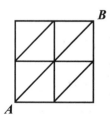

4A.
4 Minutes

Two different 3-digit numbers have the same digits, but in reverse order. No digit is zero. If the numbers are subtracted, what is the largest possible difference?

4B.
5 Minutes

A whole number is more than 50. If this number is divided by 3, the remainder is 2. If it is divided by 4, the remainder is 3. What is the smallest such whole number?

4C.
6 Minutes

The sum of 7 consecutive integers is 105. What is the sum of the least and greatest of these integers?

4D.
5 Minutes

On sale, the original price of every jacket in a store is reduced by 20%. Later, this sale price is further reduced by another 20% to create a final price. The store could have reduced the original price directly to the final price by what single percent?

4E.
7 Minutes

In this diagram, 17 toothpicks are used to form a 2-square by 3-square rectangle. How many toothpicks would be needed to form a 6-square by 8-square rectangle?

SET 13 OLYMPIAD 5

5A.
5 Minutes

When a certain tank of water is 80% empty, it contains exactly 11 gallons. How many gallons can the full tank hold?

5B.
6 Minutes

Ashley's computing machine obeys exactly two rules:
1. *If an inserted number has one digit, multiply it by 4.*
2. *If an inserted number has 2 digits, multiply the digits together.*
The first number Ashley inserts is 1. Then the result of each computation is inserted back into the machine. What is the 100th number to be inserted?

5C.
5 Minutes

As shown, the large circle contains three touching congruent circles whose centers all lie on the diameter of the large circle. The area of each small circle is 4π. If $N\pi$ represents the area of the large circle, what number does N represent?

5D.
7 Minutes

Sarah and Josh each start saving today. Sarah saves 30 cents every day. Josh saves 1 cent on the first day, 2 cents on the second day, 3 cents on the third day, and so on, saving one cent more each day than on the day before. At the end of how many days will they have saved the same total amounts?

5E.
5 Minutes

What is the smallest positive number that is divisible by 3 but not 4, by 5 but not 6, and by 7 but not 8?

1A.
3 Minutes

Find the value of the sum:

$$1^2 + 1^4 + 1^6 + 1^8 + \ldots + 1^{100}$$

1B.
5 Minutes

Jenny and Lenny pick $\frac{1}{4}$ and $\frac{1}{3}$ of a treeful of apples, respectively. Penny picks the rest of the apples. If Lenny picks 7 more apples than Jenny does, how many apples does Penny pick?

1C.
5 Minutes

The September price of a computer was $1000. Later, an October price was obtained by raising the September price by 20%. Then a November price was obtained by lowering the October price by 20%. What was the November price in dollars?

1D.
6 Minutes

In a quiz game of 20 questions, each correct answer earns 5 points. Each incorrect answer reduces the score by 2 points. Omitted questions score 0 points. Jana has a total score of 59 points. How many questions did she omit?

1E.
7 Minutes

What is the degree-measure of the angle between the hour and minute hands at 2:20 PM?

SET 14   OLYMPIAD 2

2A.
4 Minutes

What is the arithmetic mean (average) of all the positive two-digit multiples of 4?

2B.
5 Minutes

How many different five-digit palindromes are there?
(Note: A palindrome is any number that reads the same forwards and backwards, such as 151 or 729927 or 88388.)

2C.
5 Minutes

Some friends buy a video game, sharing the cost equally. If each friend pays $8, they will have $11 too much. If each friend pays $6, they will have $5 too little. What is the price of the video game in dollars?

2D.
6 Minutes

Express as a single fraction in lowest terms:

$$\frac{7}{19} \times \frac{13}{44} + \frac{7}{19} \times \frac{19}{44} + \frac{7}{19} \times \frac{25}{44} + \frac{7}{19} \times \frac{31}{44}$$

2E.
7 Minutes

Square tiles 9-inches on a side cover the floor of a rectangular room exactly. The border tiles are white and all other tiles are blue. The room measures 18 feet by 15 feet. How many tiles are white?
(Note: One foot contains 12 inches.)

3A.
3
Minutes

Express as a single number in lowest terms:

$$\frac{28 \times 26 \times 24 \times 22}{11 \times 12 \times 13 \times 14}$$

3B.
5
Minutes

The pages of a book are numbered consecutively from 1 through 177, inclusive. If a page is chosen at random, what is the probability that the page number will contain the digit "1"?

3C.
5
Minutes

Fifteen streetlights are placed at equal distances along a road. The distance between the first and third lights is 600 meters, measured between their centers. What is the distance in meters between the first and last lights?

3D.
5
Minutes

Courtney uses 9 congruent "unit" triangles to form a large triangle of 3 rows, with no gaps and no overlaps (shown). How many unit triangles does Courtney need to form a similar large triangle of 12 rows in the same way?

3E.
6
Minutes

A motorist travels 120 miles from Antwerp to Brussels at an average of 40 mph. She returns over the same road at M mph. Her average rate for the whole trip is 48 mph. What is the value of M?

SET 14 OLYMPIAD 4

4A.
4
Minutes

How many three-digit numbers satisfy all of the following?
- The sum of the tens digit and the units digit is 9;
- The number is even; and
- The number is a multiple of 3.

4B.
4
Minutes

Suppose $A \succ B$ means $\frac{A+B}{A-B}$, where A and B represent two different numbers. What is the value of $\frac{3 \succ 5}{5 \succ 3}$?

4C.
5
Minutes

As shown, each of four congruent circles just touches two other circles and two sides of the outer square. The centers of the four circles are connected to form the inner square. If the area of the outer square is 100 sq cm, what is the area of the inner square, in sq cm?

4D.
7
Minutes

In Park School's 8th grade, 33 students like volleyball, 34 like softball, 39 like basketball, 20 like volleyball and softball, 10 like volleyball and basketball, 8 like softball and basketball, 3 like all three sports, and 12 like none of these sports. How many students are in Park School's 8th grade?

4E.
5
Minutes

If a proper fraction in lowest terms is subtracted from its reciprocal, the difference is $\frac{77}{18}$. What is the proper fraction?

5A.
4
Minutes

In this multiplication example, different letters represent different digits. What digit does *H* represent in the 3-digit number *AHA*?

$$\begin{array}{r} A\,H\,A \\ \times\quad A \\ \hline T\,A\,D\,A \end{array}$$

5B.
4
Minutes

Express as a single number:

$$(-2) + (-2)(-2) + (-2)(-2)(-2) + (-2)(-2)(-2)(-2) + (-2)(-2)(-2)(-2)(-2)$$

5C.
7
Minutes

9 apes weigh as much as 4 bears. 8 bears weigh as much as 15 cougars. 10 cougars weigh as much as 27 deer. How many deer weigh the same as 4 apes?

(Note: In this problem, all members of each species weigh the same.)

5D.
7
Minutes

ABCDEF is a regular hexagon whose center is at point *O*. The area of triangle *AED* is 12 sq cm. What is the area of hexagon *ABCDEF*, in sq cm?

(Note: A hexagon is regular if all 6 sides are congruent and all 6 angles are congruent.)

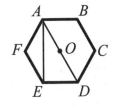

5E.
5
Minutes

If $5^{18} \times 2^{20}$ is multiplied out, what is the leading digit of the product?

(Note: The leading digit of a number is the digit in the leftmost position.)

SET 15 OLYMPIAD 1

1A.
4 Minutes

What is the first year in the 21st century (2001 through 2100) that is divisible by 11?

1B.
4 Minutes

20902 is a palindrome. What is the next larger palindrome?
(Note: A palindrome is any whole number that reads the same forwards and backwards.)

1C.
5 Minutes

The figure shown is formed by seven line segments. What is the total number of triangles in this figure?

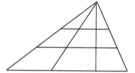

1D.
6 Minutes

On a recent Geography test, Amanda scored 5 points *below* the class average, Barb scored 8 points *above* the class average, and Colin scored 82. The average of the scores of Amanda, Barb, and Colin was equal to the class average. What was the class average?

1E.
6 Minutes

In his diary Joshua numbers all the pages consecutively, beginning with 1. This requires a total of 228 digits. How many pages are in his diary?

SET 15   OLYMPIAD 2

2A.
3
Minutes

In the correctly worked out addition problem at the right, different letters represent different digits. What digit does *A* represent?

$$\begin{array}{r} 4\,A \\ +\,A\,4 \\ \hline B\,C\,B \end{array}$$

2B.
4
Minutes

The sum of two prime numbers is 2001. What is the *larger* of these primes?

2C.
5
Minutes

It is a fact that $1 + 2 + 3 + \ldots + 28 + 29 + 30 = 465$.
What is the value of *N*, if $31 + 32 + 33 + \ldots + 58 + 59 + 60 = N$?

2D.
6
Minutes

Maxie lies on Fridays, Saturdays, and Sundays, while he tells the truth on all other days. Minnie lies on Tuesdays, Wednesdays, and Thursdays, but she is truthful on all other days. On what day of the week could they both say, "Tomorrow, I will lie."?

2E.
5
Minutes

Two congruent isosceles triangles, *ABC* and *DEF*, overlap so that their bases are parallel and the vertex of each is the midpoint of the base of the other as shown. If the area of the overlap is 12 sq cm, how many sq cm are in the area of triangle *ABC*?

3A.
4 Minutes

When Jenna opens her favorite book, the product of the page numbers on the pages facing her is 420. What is the *lesser* of the two page numbers facing Jenna?

3B.
5 Minutes

Ral, Sal, Tal, and Val each have $1.85 in quarters and dimes. No two have the same number of coins. Together, how many quarters do they have?

3C.
5 Minutes

Find the *least* whole number which can be expressed as
the sum of 2 consecutive whole numbers, AND ALSO
the sum of 3 consecutive whole numbers, AND ALSO
the sum of 5 consecutive whole numbers.

3D.
6 Minutes

Boris and Natasha start at the same place and at the same time on a 200 meter circular track and run in *opposite* directions. Boris runs at 5 meters per second and Natasha runs at 3 meters per second. In how many seconds will they first meet after starting?

3E.
7 Minutes

Four circles, each with radius 2 cm, are arranged so that each circle touches exactly two other circles at a single point as shown. If 3.14 is used to approximate π, find the area of the shaded region in sq cm, *correct to the nearest tenth.*

SET 15   OLYMPIAD 4

4A.
3 Minutes

What number is $\frac{3}{4}$ of the way from –11 to –3?

4B.
4 Minutes

The number 33,822 is divisible by each of 2, 3, 6, and 9. What is the next larger whole number also divisible by 2, 3, 6, and 9?

4C.
6 Minutes

Noelle correctly adds the lengths of <u>three</u> sides of a rectangle and gets 88 cm. Ryan correctly adds the lengths of <u>three</u> sides of the same rectangle and gets 80 cm. What is the perimeter of the rectangle in cm?

4D.
6 Minutes

Boris and Natasha start at the same place and at the same time on a 200 meter circular track and run in the *same* direction. Boris runs at 5 meters per second and Natasha runs at 3 meters per second. How many meters has Natasha run when they first meet after starting?

4E.
6 Minutes

Each diagram below shows a balance of weights using different objects. How many ⬤s will balance two △s?

5A.
4 Minutes

A quart of milk can feed either 6 cats or 10 kittens. Suppose there are 3 quarts of milk and 15 cats. After all the cats are fed, how many kittens can be fed with the leftover milk?

5B.
5 Minutes

500 students at Euclid University took a math exam. 75% of the students passed the exam. Suppose only 10% of the students had *failed* the exam. How many *more* passing grades would there have been?

5C.
6 Minutes

How many *different* sums can be obtained by adding any two different integers chosen from these thirty-five consecutive integers?

$$-17, -16, -15, \ldots, -1, 0, 1, \ldots, 15, 16, 17$$

5D.
7 Minutes

Segment $\overline{EF}$ divides rectangle $ABCD$ into square I and rectangle II. The area of rectangle $ABCD$ is 144 sq cm. The area of rectangle II is three times the area of square I. Find the *perimeter* of rectangle II, in cm.

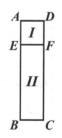

5E.
5 Minutes

Suppose 120 days ago was a Friday. What day of the week will it be 86 days from today?

1A.
4
Minutes

Find the sum of the digits in the pattern below:

```
2                 5
2 2             5 5
2 2 2         5 5 5
2 2 2 2   5 5 5 5
2 2 2 2 7 5 5 5 5
2 2 2 7 7 7 5 5 5
```

1B.
5
Minutes

Megan has three candles of the same length to provide light. Candle *A* burns for exactly 72 minutes. Candle *B* burns twice as fast as candle *A*. Candle *C* burns three times as fast as candle *B*. What is the greatest total number of minutes of light that all three candles can provide?

1C.
5
Minutes

In lowest terms, how much greater is $\dfrac{2003}{25} + 25$ than $\dfrac{2003 + 25}{25}$?

1D.
6
Minutes

Five identical circles are arranged in a straight line on a strip of paper: In how many *different* ways can exactly **three** of these circles be colored gray?
(Note: Consider the arrangements ⌐ ● ● ● ○ ○ ⌐ *and* ⌐ ○ ○ ● ● ● ⌐ *as the same because the paper can be turned around.)*

1E.
6
Minutes

The large rectangle shown is composed of five congruent smaller rectangles, each with whole number dimensions. If the perimeter of each smaller rectangle is 20 cm, find the area of the entire large rectangle in sq cm.

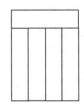

2A.
3
Minutes

Define $\triangle$ (with b top, a bottom-left, c bottom-right) as $\dfrac{a + b \times c}{a - b \times c}$.

Find the value of $\triangle$ (with 2 top, 10 bottom-left, 3 bottom-right) in simplest form.

2B.
5
Minutes

The average of five numbers is 25. Four of the numbers are 19, 21, 24, and 25. Find the other number.

2C.
5
Minutes

How many different whole numbers are factors of 100?

2D.
6
Minutes

The letters in **Twenty** and **Five** are cycled <u>separately</u> as shown and placed in a numbered vertical list. After line 1, the <u>next</u> line in which both **Twenty** and **Five** are spelled correctly is row number N. Find N.

1. Twenty   Five
2. yTwent   eFiv
3. tyTwen   veFi
4. ntyTwe   iveF
   ⋮   ⋮   ⋮
N. Twenty   Five

2E.
5
Minutes

Two squares with integer-length sides overlap so that two sides of the smaller square rest along two sides of the larger square as shown. The shaded region has area 28 sq cm. Find area of the larger square, in sq cm.

3A.
4 Minutes

A telephone call costs 25¢ for the first three minutes and 3¢ for each additional minute. If Jason pays 40¢ for a call, for how many minutes does the call last?

3B.
5 Minutes

The product of two whole numbers is 48. The average of the two numbers is 8. Find the greater of the two numbers.

3C.
5 Minutes

A famous sequence begins 1, 1, 2, 3, 5, 8, 13, 21, 34, 55, and so on. Each term of the sequence after the second term is the sum of the two terms before it. The eleventh term is 34 + 55 or 89. How many of the first thirty terms of this sequence are odd numbers?

3D.
6 Minutes

14 can be expressed as the sum of two prime numbers in exactly two different ways: 11 + 3 =14 and 7 + 7 = 14. In how many ways can 40 be expressed as the sum of two prime numbers?

3E.
6 Minutes

The area of triangle ABC is 50 sq cm. Point D lies on side $\overline{BA}$, with $\overline{DA} = 6$ cm and $\overline{BA} = 25$ cm. Find the area of triangle CBD, in sq cm.

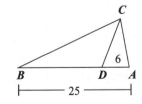

SET 16   OLYMPIAD 4

4A.
3 Minutes

HA and *AH* represent two 2-digit numbers.
If $HA - AH = 18$, what is the value of the expression $H - A$?

4B.
6 Minutes

Six darts land on a dartboard. Each dart scores 2 or 5 or 8 points. Which of the following total scores is possible?

11, 16, 25, 36, 44, 51.

4C.
5 Minutes

Ashley writes the counting numbers in four columns A, B, C, and D, as shown. If she continues in the same manner, the number 100 will appear in the column headed by which letter?

A	B	C	D
1	2	3	4
8	7	6	5
9	10	11	12
16	15	14	13

and so on ...

4D.
6 Minutes

75% of 76 is the same as 95% of what whole number?

4E.
6 Minutes

A rectangular floor, 9 ft by 11 ft, is covered completely by tiles. Each tile is either a 2 ft by 3 ft rectangle or a square 1 ft on a side. No tiles overlap. What is the least total number of tiles that could have been used to cover the floor?

5A.
3 Minutes

If $3 \times \triangle - 25 = 8$, then find the value of $3 \times \triangle + 25$.

5B.
4 Minutes

What whole number between 100 and 200 is both a perfect square and a multiple of 7?

5C.
6 Minutes

The lengths of two sides of a triangle are 6 cm and 25 cm. The third side has a length of N cm. How many whole numbers are possible values of N? *(Triangle Inequality Property: The sum of the lengths of any two sides of a triangle is greater than the length of the third side.)*

5D.
7 Minutes

Together Juan and Maria have 72 marbles. Juan gives Maria half his marbles and then 12 more marbles. Maria now has three times as many marbles as Juan has. How many marbles did Maria have originally?

5E.
5 Minutes

Each of a and b are chosen from the set of whole numbers 1, 2, 3, ..., 10. In how many different ways will $\frac{a}{b}$ have a value greater than $\frac{1}{2}$ and less than 1? For example, consider $\frac{2}{3}$ and $\frac{4}{6}$ as *two* different ways.

1A.
3
Minutes

A box of tacks weighs 120 grams when full and 70 grams when half full. How many grams does the box weigh when empty?

1B.
5
Minutes

A school's service club has six members. Two of them help in the Main Office each school day. What is the greatest number of school days that can pass without repeating the same pair of students?

1C.
5
Minutes

In the multiplication example at the right, A and B represent different digits. What 4-digit number is the product?

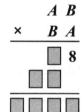

1D.
6
Minutes

A circle with radius 1 cm is inside a rectangle that is 6 cm by 10 cm. The circle rolls once around the rectangle without slipping, always touching at least one side until it returns to its starting point. Find the distance traveled by the center of the circle, in centimeters.

1E.
6
Minutes

Sadie and Rose run around a circular track in opposite directions. They each run at a constant rate and pass each other every 30 seconds. Sadie requires 45 seconds to go around the track once. How many seconds does it take for Rose to complete one lap?

SET 17   OLYMPIAD 2

2A.
3
Minutes

The product of 803 and 907 is divided by the sum of 63 and 37. What is the remainder?

2B.
5
Minutes

The average of four consecutive even integers is 17. Find the largest of the four integers.

2C.
6
Minutes

When the six-digit number $3456n7$ is divided by 8, the remainder is 5. List *both* possible values of the digit *n*.

2D.
6
Minutes

A fenced rectangular garden is 10 m wide and 20 m long. When one side is moved outward and two other sides are increased in length, the area increases by 40 sq m. What is the fewest number of meters of *additional* fencing needed to form the larger rectangular garden?

2E.
7
Minutes

The sequence 2, 3, 5, 6, 7, 10, … consists of all counting numbers which are *neither* perfect squares *nor* perfect cubes. Find the 75ᵗʰ term of this sequence.
(A perfect square is a number that can be written as the product of two equal factors. A perfect cube is a number that can be written as the product of three equal factors.)

3A.
4
Minutes

The cube shown has a different whole number from 1 through 6 written on each of its faces. The sum of the numbers on each pair of opposite faces equals 7. Find the least possible value of $A + B$.

3B.
5
Minutes

Curt mistakenly multiplied a positive number by 10, when he should have divided the original positive number by 10. The answer he found was 33.66 more than the answer he should have found. Find the original positive number.

3C.
6
Minutes

A whole number is chosen at random from the integers 11, 12, 13, …, 49, 50. What is the probability that the number is divisible by 3 or by 5?

3D.
6
Minutes

A "tower" is formed as shown by placing a small square atop a large square. The perimeter of the tower is 52 cm and the perimeter of the large square is 40 cm. Find the perimeter of the small square, in cm.

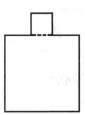

3E.
6
Minutes

Find the whole number value of n:

$$\frac{3}{80} < \frac{1}{n} < \frac{4}{101}$$

4A.
4
Minutes

What is the value of $17 \times 13 + 61 \times 13 + 22 \times 13$?

4B.
5
Minutes

An enclosed rectangular garden is 6 meters by 15 meters. Additional fencing within the garden can divide it into three smaller plots, two of them square and one rectangular. What is the least number of meters of additional fencing that is needed?

4C.
5
Minutes

The mean, the median, and the mode of the five numbers below are all equal. What number does A represent?

1.8   1.6   2.1   1.7 A

4D.
6
Minutes

List all the counting numbers less than 50 that have exactly three factors.

4E.
7
Minutes

Five years ago, Lucy's age was three times Max's age. In two years Lucy's age will be two times Max's age. How old is Max now, in years?

SET 17   OLYMPIAD 5

5A.
4 Minutes

What is the sum of all the prime numbers between 50 and 60?

5B.
4 Minutes

Adnan began with a whole number. He divided his original number by 2, subtracted 6 from the quotient, took the square root of the difference, added 1 to the square root, and took the square root of the sum. His final result was 3. What was Adnan's original number?

5C.
6 Minutes

On line segment $\overline{ABCDE}$, D is the midpoint of $\overline{AE}$. The length of $\overline{BD}$ is $\frac{2}{3}$ the length of $\overline{AB}$, and $\overline{BC} = \overline{CD}$. What percent of $\overline{AE}$ is $\overline{AC}$?

5D.
7 Minutes

Two consecutive whole numbers, each less than 20, are multiplied and their product is increased by 17. There are exactly two values of the lesser of the two consecutive numbers for which the result is *not* a prime number. Find both values of the lesser number.

5E.
5 Minutes

A digital clock shows time in the form HH:MM. On a certain day, what is the number of minutes between 7:59 AM and 2:59 PM that HH is greater than MM?

HINTS

DIVISION E
DIVISION M

HINTS: DIVISION E

SET # 1

1A. Compare each number in the first set of parentheses with its counterpart in the second set.

1B. What would happen if each number were increased by 3?

1C. How could you use a tree diagram to answer this question?

1D. Why is A a multiple of 3?

1E. Draw a stack to represent Ben's starting collection of $x + 12$ cards.

2A. Suppose Marty gave the 10 pogs to a third person, who then gave them to Jen.

2B. Draw a picture of the box made from 2-cm cubes.

2C. What are the possible values for B? How does the number of places in the partial products affect the choices?

2D. A remainder of 4 means that each answer must be a divisor of what number?

2E. Suppose only adults buy tickets.

3A. What time is it 24 hours from now? 48 hours?

3B. How many cubes are in each stack, or in each layer?

3C. How many games did the Panthers lose?

3D. Suppose there are three more students. How many students would be left over then?

3E. Draw a diagram, letting a box represent one-third of the number.

4A. What does $\frac{4}{5}$ of 100 mean?

4B. Find the area of one square. How many different ways can you find to solve this problem?

4C. Make an organized list of factors, pairing each with its cofactor.

4D. How far will the car travel in 3 minutes?

4E. How much should a purchase of 5 pens and 5 pencils cost?

5A. A terminal zero is produced by multiplying any even number by 5.

5B. Don't forget the squares that are tilted.

5C. What is the largest number of quarters possible?

5D. Draw the clock.

5E. Work backwards.

HINTS: DIVISION E

SET # 2

1A. Group the numbers.
1B. Decide whether to view the stairway from the front, side, or top.
1C. Compute each purchase separately.
1D. A Venn diagram can help.
1E. How many "extra" pounds does Todd weigh? By how many pounds does this raise the average?

2A. What number does Ana start with?
2B. Represent Jay's collection by a box and the other people's collections by stacks of boxes.
2C. What value must be entered in the top left square?
2D. Combine parts of squares.
2E. Act it out, starting with the simplest cases.

3A. Divide both sides by each common factor.
3B. Start by listing appropriate multiples of 5.
3C. Count up from 7:45, using round numbers to keep it simple.
3D. Find both perimeters.
3E. Start with the simplest cases and look for a pattern.

4A. What kind of numbers have exactly two factors?
4B. Since 6 cm represents 15 feet, what does 10 cm represent?
4C. Examine several simpler cases and look for a pattern.
4D. Suppose Marc makes his purchase three times.
4E. Replace the box, trying several numbers.

5A. When folded into a cube, which face will not touch the face with *T*?
5B. Place each age on a "number" line..
5C. Which subtraction produces the 7 in 3497?
5D. What are the outer dimensions of the overall figure?
5E. Consider all multiples of each of the denominators.

SET # 3

1A. Combine the numbers to the left of the equals sign.
1B. Place the middle of the 9 numbers in the middle circle.
1C. How many pages does she read every week?
1D. What is the smallest number divisible by each of 2, 3, 4, 5, and 6?
1E. How much larger is the area of the whole circle than that of the whole triangle?

2A. How many 123s are there?
2B. Why does Lee walk 7 m more than Marina?
2C. Find the number of minutes each needs to drive 20 miles.
2D. How can you combine sales of the five circles and the two circles to produce a sale involving twelve circles?
2E. Suppose each dart scores 5 points. Then find a way to trade 5-point darts for other darts.

3A. Start with four 9s and adjust their ages to fit the facts.
3B. A multiple of 27 is also a multiple of what other number?
3C. Generate the first several numbers until a pattern emerges.
3D. List all multiples of 6 less than 100. Then check for divisibility by 5 and by 4.
3E. Peel away each outer layer.

4A. Divide 2000 by 37.
4B. Find Q first.
4C. Compare two tables of possible values: actual votes cast to hypothetical votes cast.
4D. Build up from the smallest stone.
4E. Draw a tree diagram to list all paths.

5A. How many more figures did Allen have than Charlie at first?
5B. Compare each page number to that of the middle page.
5C. How large is each of the equal intervals (jumps) from 9 to 21?
5D. Assign some values to DDD and look for the largest common prime factor.
5E. How many points were scored altogether? How many events are possible?

HINTS: DIVISION E

SET # 4

1A. List two- and three-digit numbers separately.
1B. Make a table of all possible lengths and widths.
1C. How can you use the LCM of 3 and 4?
1D. Look at the figure, one layer at a time.
1E. What happens if you remove 2 lame ducks and 4 sitting ducks?

2A. Examine each score separately to see if it is possible.
2B. How much did each item cost?
2C. Draw the measuring stick. The interval is divided into how many equal parts?
2D. Examine the faces from each of the six directions.
2E. Because of the 3¢ stamp, check each amount until three consecutive values can be made.

3A. Examine the ones, tens and hundreds places separately.
3B. Work backwards. Where should you start?
3C. List all products and cross out duplicates.
3D. Make a table of all pairs whose product is 36.
3E. Act it out, one traffic light at a time.

4A. Put the largest digits in the tens place.
4B. Select the highest available disks for Gina.
4C. How many minutes does the fast clock gain each hour?
4D. Compare the last value in each row to the row number.
4E. Write out the number and divide until a pattern emerges.

5A. Add from right to left, replacing letters in order.
5B. Since $A + B = 10$, list all possible 2-digit numbers AB.
5C. Compute the missing lengths first.
5D. How many multiples of 9 are there? of 6? How many duplications?
5E. Each terminal zero is the product of 5 and 2.

HINTS: DIVISION E

SET # 5

1A. What is the cost of 1 letter? 2? 3? 4?
1B. Start with the last two facts.
1C. What happens if 5 is added to the counting number?
1D. What is the area of one of the squares?
1E. Are any chairs counted twice?

2A Work backwards.
2B. Act it out.
2C. List all possibilities, considering one sentence at a time
2D. How does the width of the rectangle compare to its length?
2E. How far apart are the two cars at 1:30 PM? 2:00 PM? 2:30 PM?

3A. What do three widgets cost?
3B. What is the difference in their ages?
3C. How many minutes are lost between 1 PM one day and 10 AM the next?
3D. What is the area of the square? the rectangle?
3E. How many numbers are in the set?

4A. Fill in each box from right to left.
4B. Compute separately the expenses and the income.
4C. Finding the factors of 42 may help.
4D. What digits could *B* represent?
4E. Represent the ladder as a number line and the middle rung as zero.

5A. How much did Tony pay for the two extra candy bars?
5B. How many books were bound every two days?
5C. Use place value to decide which letter represents 9.
5D. Draw a diagram.
5E. Draw a time line. How many intervals are there between classes?

HINTS: DIVISION E

SET # 6

1A. Group the numbers to subtract similar numbers first.
1B. Act out the placing of the marbles into the boxes.
1C. Consider one condition at a time.
1D. Start with a table of postage amounts using only 1-cent and 5-cent stamps.
1E. What is the length of one side of a square?

2A. What is the sum of the digits?
2B. Change this problem into an addition.
2C. Act it out.
2D. What is the total surface area of the cube?
2E. How much did each new member pay?

3A. Consider one condition at a time.
3B. Originally, the staircase contains how many cubes?
3C. Make an organized list.
3D. What is the first time after 1 PM that both lights flash together?
3E. How many pennies might be possible?

4A. Seven hours from now, what numeral will the hand point to?
4B. What is the length of one side of the large square?
4C. What is the sum of the six numbers? the seven numbers?
4D. What number should occupy the box above *B*?
4E. Work backwards.

5A. How many stamps does he have at the end of the fourth day? the fifth?
5B. Does an answer exist before 63,333?
5C. Compare the sums using just two circles at a time.
5D. In an organized way, count separately the exterior and interior squares.
5E. Keep the sum constant at 44 and try different differences.

HINTS: DIVISION E

SET # 7

1A. Which numbers are the easiest to multiply?

1B. How could removing the two people at the ends of the tables simplify the solution?

1C. 24 represents how many girls per boy?

1D. What numbers are multiples of two, but not of all three, of the numbers?

1E. Sketch the diagrams for each possible seating.

2A. How many pencils are in one bundle?

2B. How many barrels of oil are on hand?

2C. In *ABCD*, what digit does *A* represent?

2D. How many multiples of 7 are less than 100? less than 1000?

2E. What are the dimensions of the whole figure?

3A. Start with the last condition and list the possibilities.

3B. How many choices are there for the hundreds digit? tens digit? ones digit?

3C. What is the value of 100 – 98? 96 – 94? 92 – 90?

3D. Draw diagrams showing the results of one cut, two cuts, three cuts.

3E. The imaginary arrangement has 15 rows of 6 extra seats per row. Where would these extra seats come from?

4A. Meg is how many years older than Beth?

4B. Use place value to decide which letter should represent 3.

4C. Can I pick 4 gumballs and not have any two match in color?

4D. Make a list of all possible lengths and widths.

4E. Suppose all 70 vehicles are motorcycles. How many more of the 269 tires are not accounted for?

5A. Work backwards.

5B. Draw a time line. Label one point for "today".

5C. How many laughs and how many smiles equal 30 grins?

5D. Find the height, length, and width of the box.

5E. First find the digit *D*. Then apply a test of divisibility for 3.

HINTS: DIVISION E

SET # 8

1A. If two equal quantities are each divided by the same number, are the results equal?
1B. How does the hundreds digit compare to the sum of all three digits?
1C. For every $3 that Marisa spends, how much *more* does Andie spend than Marisa?
1D. What is the area of the rectangle?
1E. What are the first few common multiples of 9 and 6?

2A. What is the average of the five numbers?
2B. Draw diagrams for each possibility.
2C. How many tablets are in the bottle?
2D. What is the sum of one length and one width?
2E. How many times is 4 written in the ones place? the tens place?

3A. After the starting number is multiplied by 4, what would happen if the starting number is added twice to the result?
3B. How many games did Hannah play?
3C. What is the average number of letters that each house gets?
3D. What are the dimensions of the smaller rectangles?
3E. Make an organized list, starting with zero.

4A. If Madison buys 10 five-cent stamps, how much of the $1.00 is not accounted for?
4B. What must be true of the tens digit?
4C. Describe the new set if each number in the given set is increased by 1.
4D. The resulting solid has how many faces? What is the area of each face?
4E. Each of the numbers will divide 42.

5A. Suppose one cat distributes her boots to the other three cats.
5B. Compare the multiples of 6 and 5.
5C. List each possibility.
5D. When folded into the carton, which sides are the same length?
5E. What digit does *B* represent?

HINTS: DIVISION E

SET # 9

1A. Is anyone counted twice?
1B. Name each girl and then list all pairs.
1C. What is the result of adding all three purchases?
1D. What is the sum of the length and the width?
1E. Group conveniently by place value.

2A. How much of the wall across is not covered by the picture?
2B. Work backwards.
2C. Test various values for Carlos' age.
2D. Examine the chart two rows at a time.
2E. How many cubes do not touch the box?

3A. In all, how many T-shirts were sold?
3B. Which numbers should be eliminated from the list?
3C. Use a Venn diagram.
3D. What is the least number divisible by all of the denominators?
3E. How long are each of the segments?

4A. Start with the smallest pile.
4B. Can the sum of two odd numbers be an odd number?
4C. In simplest terms, what is the ratio of children to adults?
4D. Which segments have equal lengths?
4E. What is the total of all 80 numbers?

5A. What other days during May 2025 occur on Sundays?
5B. What is the test of divisibility for 9?
5C. Make a table listing all ways to score 25 points in 8 turns.
5D. How can two Els be packed so as to occupy the least amount of space?
5E. Make and compare two lists.

HINTS: DIVISION E

SET # 10

1A. Pair "similar" numbers.
1B. Which face will touch all of the other faces?
1C. Who is the flute player?
1D. How many toothpicks must be added to make one more triangle?
1E. The overall average rate is the total distance divided by the total time.

2A. Any number divisible by both 3 and 5 is also divisible by what other number?
2B. What is the difference now? next week? the week after?
2C. What is the average of everyone's scores for the 3 races?
2D. What are the sums now? What changes could make the three row-sums equal?
2E. Start in any direction.

3A. Draw each possibility.
3B. What would Jeffrey earn if he did no additional chores?
3C. Use tests of divisibility.
3D. Split the figure into a rectangle and a triangle.
3E. How much time does the faster watch gain on the slower watch each hour?

4A. What is the first multiple of 25 greater than 259?
4B. What are the factors of 6?
4C. Additions are usually easier than subtractions. Change the problem.
4D. Split the square into 9 congruent smaller squares.
4E. Count up in an organized way.

5A. What is the product of 6 and 150?
5B. Create a "super-stamp" consisting of one stamp of each type.
5C. What is the total length? width?
5D. What is the sum of the numbers from 1 through 9 inclusive?
5E. What is the probability of drawing a black marble?

SET # 11

1A. Which two numbers should you start with?

1B. Why is it safe to average only the largest and smallest multiples of 3?

1C. Make an organized list.

1D. What are the prime factors of 45?

1E. Rotate the lower square about the center of the upper square.

2A. How long is each segment of side *DC*?

2B. What are the first three row numbers for the correct spelling of MATH? of OLYMPIADS?

2C. Are all prime numbers odd?

2D. Only large divisors can produce large remainders.

2E. If you draw an altitude from *E* to side *BC*, which regions will have the same area?

3A. Make two lists and compare.

3B. Which multiples of 16 end in a 2?

3C. How many students does each teacher teach each day?

3D. How many times does the sum $0 + 1 + 2 + \ldots + 9$ appear?

3E. Simplify from the bottom up, starting with $4 + \frac{3}{4}$.

4A. How many more male children than female children are in the family?

4B. Work backwards.

4C. Use a Venn diagram.

4D. Three of the numbers are 4, 9, and 25.

4E. Trace the path of the center of the circle.

5A. There are as many different ways to choose three points as to omit the other two points.

5B. What is the total number of votes for *B*, *C*, *D*, and *E*?

5C. Any number divisible by 99 is also divisible by 9 and 11.

5D. How many desserts can 1 chef prepare in 10 minutes? 15 minutes?

5E. Sketch the region reachable by the goat, then partition it into quarter-circles.

SET # 12

1A. Start with the divisibility.

1B. What is the area of the two larger squares?

1C. Evaluate the first expression first.

1D. What is the total of all his tests? the first three tests?

1E. How much would she receive for the last two weeks? each week?

2A. What are the factor pairs of 96?

2B. What is the units digit in the multiplicand?

2C. What is the width of the shaded rectangle?

2D. What is the smallest total possible? the largest?

2E. What are the prime factors of a number that ends in a zero?

3A. Angle *BAE* is acute.

3B. Start by finding the product of the possible digits.

3C. If $a = b + c$, then $c = a - b$, where a, b, and c are any numbers.

3D. How long does it take to lose one hour? How many hours must it lose before it is again correct?

3E. How many paths are there from *O* to *L*? *L* to *Y*? *O* to *Y*?

4A. Work backwards.

4B. Divide repeatedly by 2.

4C. Compare the entries in every other row.

4D. First solve for *D*.

4E. What is the greatest number of intersections for 2 chords? 3? 4?

5A. List each decade of the set separately.

5B. What are the common factors of 35 and 42?

5C. Which 10 slips does each draw? Are there common differences?

5D. How much additional money was collected?

5E. Replace *A* by possible values and look for the common factors.

HINTS: DIVISION M

SET # 13

1A. What month will it be in 12 months? 24? 36?
1B. What is the total for everyone except Barry?
1C. What percent of the mathletes are in one or both categories?
1D. What is the largest number of cancellations possible within one problem?
1E. Organize the triangles by the number of regions each contains.

2A. How many marbles can I pick before I am forced to pick a red?
2B. What percent of their matches did Emily win?
2C. The five-digit number is divisible by both 3 and 4.
2D. Find an equivalent fraction to avoid dividing by a decimal.
2E. How long is one side of the large square?

3A. What is the value of $1 - 3$? $5 - 7$?
3B. Compare Chris's purchase to Dan's.
3C. What is the value of E? What must be true of $U \times M$ to produce this result?
3D. How many minutes and how many miles are there between the first two mile markers?
3E. Starting from A, how many ways are there to reach the center of the figure?

4A. What must be the hundreds digit of the two three-digit numbers?
4B. Suppose 1 is added to the number. What would happen if you then divide by 3 and also by 4?
4C. What is the middle integer (median) of the set?
4D. Choose a convenient original price for a jacket and then apply the discounts.
4E. Count the horizontal and vertical toothpicks separately.

5A. What percent full is the tank?
5B. Start with 1 and follow the rules until you see a pattern.
5C. What is the radius of each small circle? the large circle?
5D. For how many days does Sarah save more than Josh? the same amount? less?
5E. What set of numbers are divisible by 3, 5, and 7?

HINTS: DIVISION M

SET # 14

1A. What is the value of 1 raised to any power?

1B. What is the difference between $\frac{1}{3}$ and $\frac{1}{4}$?

1C. The November price is different from the September price.

1D. To score 59 points, what is the fewest number of correct answers possible?

1E. Through how many degrees does the hour hand move in 5 minutes?

2A. Average the largest and smallest multiples of 4. Why does this work?

2B. How are the last two digits related to the first two digits?

2C. What are the differences in per-person cost and in total amount collected?

2D. Use the distributive property and then simplify.

2E. How many tiles border two edges of the floor?

3A. Use the commutative property and then separate the fractions.

3B. How many 1s appear in the ones place? tens? hundreds?

3C. There are not 15 distances between the first and last light.

3D. How many unit triangles are needed to form 1 row? 2 rows? 3 rows?

3E. The average rate equals the total distance divided by the total time.

4A. Consider one condition at a time.

4B. Substitute and simplify.

4C. How does a diameter of a circle compare to a side of the outer square.

4D. Use a Venn diagram.

4E. The denominators of the two fractions are both factors of 18.

5A. For what values of A does $A \times A$ produce a units digit of A?

5B. In what order should the operations be done?

5C. How many apes weigh the same as 8 bears?

5D. What is the area of triangle OED? How many such triangles form the hexagon?

5E. The product of 5 and 2 ends in a zero.

HINTS: DIVISION M

SET # 15

1A. Use the test of divisibility for 11.
1B. How do the first two digits affect the last two digits?
1C. Count the number of triangles layer by layer.
1D. How does Colin's score compare to the class average?
1E. How many pages have 1 digit? 2? 3?

2A. First find the value of *B*.
2B. Are all primes odd?
2C. How much is added to each number of the first sequence to get the corresponding number of the second sequence?
2D. Which days are possible for Maxie? Minnie?
2E. Draw a convenient line.

3A. What is the ones digit for one of the page numbers?
3B. How many different ways can you combine quarters and dimes to make $1.85?
3C. Begin with sets of five consecutive whole numbers.
3D. How far apart are they after 1 second? 2? 3?
3E. Connect the centers of the circles.

4A. How far is it from −11 to −3?
4B. What is the smallest number divisible by 2, 3, 6, and 9?
4C. Together, Noelle and Ryan have added how many lengths and how many widths?
4D. How far apart are they after 1 second? 2? 3?
4E. Add two circles to each balance pan in the first diagram.

5A. How much milk is needed to feed 15 cats?
5B. In each case how many students pass?
5C. What is the least possible sum? the greatest?
5D. Draw two lines to split rectangle *EBCF* into 3 congruent figures.
5E. How many days elapsed after that Friday?

HINTS: DIVISION M

SET # 16

1A. Can you see two overlapping number triangles?
1B. For how many minutes does candle *B* burn? candle *C*?
1C. Can you form two separate fractions from the second given fraction?
1D. List all arrangements in an orderly way.
1E. The length of a smaller rectangle is how many times its width?

2A. Substitute and simplify.
2B. What is the sum of the five numbers?
2C. Pair each factor with its cofactor.
2D. What are the row numbers for the first three correct spellings of TWENTY? of FIVE?
2E. The shaded area is the difference between the areas of the two squares.

3A. How much does the call cost after the first three minutes?
3B. If the average of two numbers is 8, what is their sum?
3C. Write only the ones digits for the first 30 terms.
3D. Use the smaller primes to generate a list of addend pairs.
3E. What is the length of the altitude to side *BA*?

4A. Assign several pairs of values to *H* and *A* to find a pattern.
4B. Start with each dart scoring 5 points and then move darts to the 2- or 8- region.
4C. Compare corresponding values in every other row.
4D. What is 75% of 76?
4E. The more large tiles used, the fewer small tiles needed.

5A. How much larger is $3 \times \triangle + 25$ than $3 \times \triangle - 25$?
5B. List all perfect squares between 100 and 200.
5C. Use the Triangle Inequality Property.
5D. Work backwards.
5E. Make an organized list.

HINTS: DIVISION M

SET # 17

1A. How many grams of tacks are removed?
1B. Represent the students by A, B, C, ..., F and organize a list of all member pairs.
1C. The units digit of the product $B \times A$ is 8. Examine all possible values of A and B.
1D. Draw the picture.
1E. What part of the track does Sadie cover between meetings?

2A. First find the sum of 63 and 37.
2B. What four consecutive even numbers surround 17?
2C. What six-digit number is divisible by 8?
2D. Draw both possible rectangles to show the extensions.
2E. In the first 75 or 80 counting numbers, which numbers do not belong?

3A. What numbers are not possible values for A and B?
3B. Estimate Curt's probable answer.
3C. List all multiples of 3 or 5 in the interval.
3D. Where does the extra 12 cm in the perimeter of the tower come from?
3E. What is the least common *numerator*?

4A. 13 is the common factor in each of the three terms.
4B. Draw both possible diagrams.
4C. Start by finding the mean of the four given numbers.
4D. Which of the first ten counting numbers has exactly three factors? Then test some others.
4E. Both now and in 7 years, Lucy's age will be a multiple of Max's age.

5A. Use tests of divisibility on each possible number.
5B. Work backwards
5C. Draw the diagram.
5D. 17 more than a multiple of 17 is another multiple of 17.
5E. List the times hour by hour.

ANSWERS

DIVISION E
DIVISION M

ANSWERS: DIVISION E

SET # 1

Olympiad 1	Olympiad 2	Olympiad 3	Olympiad 4	Olympiad 5
1A. 16	**2A.** 14	**3A.** 2:45	**4A.** 20	**5A.** 6
1B. 197	**2B.** 6	**3B.** 23	**4B.** 9	**5B.** 10
1C. 10	**2C.** 299	**3C.** 5	**4C.** 9	**5C.** 4
1D. 6	**2D.** 6, 9 or 18	**3D.** 17	**4D.** 40	**5D.** 3 and 8
1E. 36	**2E.** 5	**3E.** 6	**4E.** 17	**5E.** 27

SET # 2

Olympiad 1	Olympiad 2	Olympiad 3	Olympiad 4	Olympiad 5
1A. 500	**2A.** 16	**3A.** 7	**4A.** 6	**5A.** M
1B. 50	**2B.** 8	**3B.** 135	**4B.** 80	**5B.** 6
1C. Gameland, 5	**2C.** 2	**3C.** 8 hrs 20 min	**4C.** 2	**5C.** 9346
1D. 8	**2D.** 18	**3D.** 32	**4D.** 8	**5D.** 54
1E. 5	**2E.** 6	**3E.** 400	**4E.** 3 and 4	**5E.** 12

SET # 3

Olympiad 1	Olympiad 2	Olympiad 3	Olympiad 4	Olympiad 5
1A. 29	**2A.** 2460	**3A.** 7	**4A.** 1998	**5A.** 3
1B. 8	**2B.** 46	**3B.** 5	**4B.** 25	**5B.** 42
1C. Saturday	**2C.** 6	**3C.** 16	**4C.** 30	**5C.** 4001
1D. 5	**2D.** 61	**3D.** 54, 55, and 56	**4D.** 5	**5D.** 555 or 777
1E. 7	**2E.** 8	**3E.** 27	**4E.** 28	**5E.** 5

SET # 4

Olympiad 1	Olympiad 2	Olympiad 3	Olympiad 4	Olympiad 5
1A. 9	**2A.** 6, 17 and 58	**3A.** 20	**4A.** 8352	**5A.** 3, 7, 8 and 5
1B. 42	**2B.** 54	**3B.** 60	**4B.** 31	**5B.** 82 and 28
1C. 120	**2C.** 39	**3C.** 19	**4C.** 2:46 PM	**5C.** 99
1D. 100	**2D.** 26	**3D.** 4	**4D.** 145	**5D.** 44
1E. 37	**2E.** 13	**3E.** 50	**4E.** 4	**5E.** 7

SET # 5

Olympiad 1	Olympiad 2	Olympiad 3	Olympiad 4	Olympiad 5
1A. 66	**2A.** 15	**3A.** 13	**4A.** 9	**5A.** 95
1B. Krisha: 10 AM,	**2B.** Half-dollar	**3B.** 11	**4B.** 2	**5B.** 8
Lisa: noon,	**2C.** 2149	**3C.** 8:57	**4C.** 3	**5C.** 2056
Jean: 2 PM	**2D.** 64	**3D.** 30	**4D.** A = 7, B = 5	**5D.** 2500
1C. 61	**2E.** 4:30	**3E.** 235	**4E.** 23	**5E.** 42
1D. 56				
1E. 600				

SET # 6

Olympiad 1	Olympiad 2	Olympiad 3	Olympiad 4	Olympiad 5
1A. 300	**2A.** 357	**3A.** 1926	**4A.** 1	**5A.** 78
1B. 29	**2B.** 6754	**3B.** 50	**4B.** 22	**5B.** 444
1C. 45	**2C.** 5	**3C.** 20	**4C.** 11	**5C.** 16
1D. 79	**2D.** 300	**3D.** 3:06	**4D.** A = 2, B = 1	**5D.** 72
1E. 180	**2E.** 200	**3E.** 8	**4E.** 38	**5E.** 32

SET # 7

Olympiad 1	Olympiad 2	Olympiad 3	Olympiad 4	Olympiad 5
1A. 34,000	**2A.** 720	**3A.** 84	**4A.** 5	**5A.** 8
1B. 10	**2B.** 3	**3B.** 27	**4B.** 10,599	**5B.** Sunday
1C. 8	**2C.** 1089	**3C.** 50	**4C.** 45	**5C.** 10
1D. 3	**2D.** 128	**3D.** 7	**4D.** 74	**5D.** 112
1E. Amy - Red,	**2E.** 64	**3E.** 540	**4E.** 54	**5E.** 12,453
Dawn - Blue,				
Soumiya - Green				

SET # 8

Olympiad 1	Olympiad 2	Olympiad 3	Olympiad 4	Olympiad 5
1A. 50	**2A.** 7	**3A.** 7	**4A.** 6	**5A.** 60
1B. 651	**2B.** 6	**3B.** 2	**4B.** 25	**5B.** 26
1C. 300	**2C.** 21	**3C.** 6	**4C.** 25	**5C.** 24
1D. 6	**2D.** 6, 10, or 12	**3D.** 40	**4D.** 122	**5D.** 222
1E. 72	**2E.** 40	**3E.** 13	**4E.** 5	**5E.** 1

ANSWERS: DIVISION E

SET # 9

Olympiad 1	Olympiad 2	Olympiad 3	Olympiad 4	Olympiad 5
1A. 36	**2A.** 11	**3A.** 200	**4A.** 21	**5A.** Thursday
1B. 10	**2B.** 22	**3B.** 124	**4B.** 2	**5B.** 6
1C. 30	**2C.** 6	**3C.** 10	**4C.** 4	**5C.** 7
1D. 350	**2D.** T	**3D.** 24	**4D.** 5	**5D.** 6
1E. 175	**2E.** 52	**3E.** 72	**4E.** 50	**5E.** 25

SET # 10

Olympiad 1	Olympiad 2	Olympiad 3	Olympiad 4	Olympiad 5
1A. 500	**2A.** 120	**3A.** DBAC	**4A.** 16	**5A.** 30
1B. W	**2B.** 8	**3B.** 3	**4B.** $\frac{4}{6}$ or $\frac{2}{3}$	**5B.** 6
1C. Alexis	**2C.** 11	**3C.** 8640	**4C.** 5438	**5C.** 42
1D. 44	**2D.** 4 and 11	**3D.** 36	**4D.** 63	**5D.** 1 or 3
1E. 36	**2E.** 8	**3E.** 1:00 PM	**4E.** 55	**5E.** 16

ANSWERS: DIVISION M

SET # 11

Olympiad 1	Olympiad 2	Olympiad 3	Olympiad 4	Olympiad 5
1A. 14	2A. 68	3A. 53	4A. 5	5A. 10
1B. 51	2B. 37	3B. 7	4B. 13	5B. 13
1C. 7	2C. 19	3C. 80	4C. 7	5C. 46
1D. 75	2D. 15	3D. 901	4D. 5	5D. 10
1E. 25	2E. 48	3E. $\frac{57}{88}$	4E. 22.28	5E. 50

SET # 12

Olympiad 1	Olympiad 2	Olympiad 3	Olympiad 4	Olympiad 5
1A. 3715	2A. Apr. 24	3A. 10	4A. 10	5A. 45
1B. 18	June 16,	3B. 12 and 56	4B. 583	5B. 12
1C. 9	Aug. 12, and	3C. 30	4C. F	5C. 70
1D. 90	Dec. 8,	3D. 60	4D. 647	5D. 18
1E. 1200	2B. 2	3E. 48	4E. 15	5E. 9
	2C. 33			
	2D. 48			
	2E. 15			

SET # 13

Olympiad 1	Olympiad 2	Olympiad 3	Olympiad 4	Olympiad 5
1A. March	2A. 33	3A. –10	4A. 792	5A. 55
1B. 35	2B. 3	3B. 50	4B. 59	5B. 6
1C. 30	2C. 5	3C. 625	4C. 30	5C. 36
1D. $\frac{1}{9}$	2D. 25	3D. 9:46	4D. 36	5D. 59
1E. 16	2E. 42	3E. 13	4E. 110	5E. 105

SET # 14

Olympiad 1	Olympiad 2	Olympiad 3	Olympiad 4	Olympiad 5
1A. 50	2A. 54	3A. 16	4A. 15	5A. 1
1B. 35	2B. 900	3B. $\frac{97}{177}$	4B. –1	5B. –22
1C. 960	2C. 53	3C. 4200	4C. 25	5C. 9
1D. 4	2D. $\frac{14}{19}$	3D. 144	4D. 83	5D. 36
1E. 50	2E. 84	3E. 60	4E. $\frac{2}{9}$	5E. 4

ANSWERS: DIVISION M

SET # 15

Olympiad 1	Olympiad 2	Olympiad 3	Olympiad 4	Olympiad 5
1A. 2002	2A. 7	3A. 20	4A. −5	5A. 5
1B. 21012	2B. 1999	3B. 16	4B. 33,840	5B. 75
1C. 18	2C. 1365	3C. 15	4C. 112	5C. 67
1D. 85	2D. Thursday	3D. 25	4D. 300	5D. 48
1E. 112	2E. 24	3E. 3.4	4E. 7	5E. Monday

SET # 16

Olympiad 1	Olympiad 2	Olympiad 3	Olympiad 4	Olympiad 5
1A. 147	2A. 4	3A. 8	4A. 2	5A. 58
1B. 120	2B. 36	3B. 12	4B. 36	5B. 196
1C. 24	2C. 9	3C. 20	4C. D	5C. 11
1D. 6	2D. 13	3D. 3	4D. 60	5D. 12
1E. 80	2E. 64	3E. 38	4E. 19	5E. 20

SET # 17

Olympiad 1	Olympiad 2	Olympiad 3	Olympiad 4	Olympiad 5
1A. 20	2A. 21	3A. 5	4A. 1300	5A. 112
1B. 15	2B. 20	3B. 3.4	4B. 9	5B. 140
1C. 1008	2C. 3 and 7	3C. $\frac{9}{20}$	4C. 1.8	5C. 40
1D. 24	2D. 4	3D. 24	4D. 4, 9, 25, and 49	5D. 16 and 17
1E. 90	2E. 86	3E. 26	4E. 12	5E. 53

PERCENTS CORRECT

CORRECT

DIVISION E
DIVISION M

% CORRECT: DIVISION E

The scoring reported by the PICOs following each contest resulted in the percents below. To some extent, each percent indicates how hard those students found that problem at that time. The following table interprets the level of difficulty for a given problem and may help guide you when selecting problems.

Percent Correct	Assumed Level of Difficulty
0% – 19%	Challenging
20% – 49%	Moderate and typical
50% – 100%	Straightforward

SET 1

Olympiad 1	Olympiad 2	Olympiad 3	Olympiad 4	Olympiad 5
1A. 11%	2A. 42%	3A. 53%	4A. 37%	5A. 51%
1B. 44%	2B. 37%	3B. 41%	4B. 66%	5B. 58%
1C. 65%	2C. 38%	3C. 34%	4C. 46%	5C. 48%
1D. 22%	2D. 24%	3D. 56%	4D. 36%	5D. 61%
1E. 31%	2E. 60%	3E. 33%	4E. 27%	5E. 17%

SET 2

Olympiad 1	Olympiad 2	Olympiad 3	Olympiad 4	Olympiad 5
1A. 80%	2A. 35%	3A. 54%	4A. 39%	5A. 78%
1B. 79%	2B. 51%	3B. 66%	4B. 14%	5B. 60%
1C. 59%	2C. 50%	3C. 50%	4C. 9%	5C. 44%
1D. 30%	2D. 44%	3D. 15%	4D. 55%	5D. 13%
1E. 23%	2E. 38%	3E. 26%	4E. 40%	5E. 32%

SET 3

Olympiad 1	Olympiad 2	Olympiad 3	Olympiad 4	Olympiad 5
1A. 74%	2A. 78%	3A. 14%	4A. 39%	5A. 56%
1B. 33%	2B. 26%	3B. 15%	4B. 30%	5B. 39%
1C. 33%	2C. 7%	3C. 23%	4C. 13%	5C. 8%
1D. 8%	2D. 56%	3D. 48%	4D. 7%	5D. 4%
1E. 10%	2E. 7%	3E. 7%	4E. 1%	5E. 11%

% CORRECT: DIVISION E

SET 4

Olympiad 1	Olympiad 2	Olympiad 3	Olympiad 4	Olympiad 5
1A. 28%	**2A.** 49%	**3A.** 43%	**4A.** 31%	**5A.** 67%
1B. 17%	**2B.** 40%	**3B.** 76%	**4B.** 44%	**5B.** 58%
1C. 5%	**2C.** 22%	**3C.** 5%	**4C.** 13%	**5C.** 10%
1D. 58%	**2D.** 24%	**3D.** 29%	**4D.** 25%	**5D.** 11%
1E. 52%	**2E.** 4%	**3E.** 23%	**4E.** 46%	**5E.** 4%

SET 5

Olympiad 1	Olympiad 2	Olympiad 3	Olympiad 4	Olympiad 5
1A. 42%	**2A.** 66%	**3A.** 46%	**4A.** 45%	**5A.** 54%
1B. 78%	**2B.** 49%	**3B.** 58%	**4B.** 29%	**5B.** 46%
1C. 31%	**2C.** 47%	**3C.** 23%	**4C.** 18%	**5C.** 11%
1D. 17%	**2D.** 14%	**3D.** 35%	**4D.** 42%	**5D.** 13%
1E. 25%	**2E.** 11%	**3E.** 5%	**4E.** 26%	**5E.** 17%

SET 6

Olympiad 1	Olympiad 2	Olympiad 3	Olympiad 4	Olympiad 5
1A. 64%	**2A.** 46%	**3A.** 70%	**4A.** 83%	**5A.** 58%
1B. 53%	**2B.** 43%	**3B.** 45%	**4B.** 14%	**5B.** 20%
1C. 49%	**2C.** 41%	**3C.** 26%	**4C.** 39%	**5C.** 41%
1D. 3%	**2D.** 15%	**3D.** 27%	**4D.** 53%	**5D.** 16%
1E. 17%	**2E.** 16%	**3E.** 59%	**4E.** 13%	**5E.** 57%

SET 7

Olympiad 1	Olympiad 2	Olympiad 3	Olympiad 4	Olympiad 5
1A. 53%	**2A.** 50%	**3A.** 61%	**4A.** 57%	**5A.** 84%
1B. 65%	**2B.** 44%	**3B.** 18%	**4B.** 34%	**5B.** 38%
1C. 25%	**2C.** 13%	**3C.** 13%	**4C.** 25%	**5C.** 12%
1D. 21%	**2D.** 10%	**3D.** 26%	**4D.** 18%	**5D.** 12%
1E. 56%	**2E.** 12%	**3E.** 18%	**4E.** 12%	**5E.** 37%

% CORRECT: DIVISION E

SET 8

Olympiad 1	Olympiad 2	Olympiad 3	Olympiad 4	Olympiad 5
1A. 65%	2A. 48%	3A. 64%	4A. 77%	5A. 64%
1B. 66%	2B. 42%	3B. 65%	4B. 34%	5B. 63%
1C. 23%	2C. 33%	3C. 25%	4C. 52%	5C. 29%
1D. 49%	2D. 27%	3D. 27%	4D. 23%	5D. 8%
1E. 31%	2E. 18%	3E. 29%	4E. 14%	5E. 56%

SET 9

Olympiad 1	Olympiad 2	Olympiad 3	Olympiad 4	Olympiad 5
1A. 59%	2A. 58%	3A. 76%	4A. 70%	5A. 70%
1B. 49%	2B. 45%	3B. 47%	4B. 53%	5B. 62%
1C. 40%	2C. 74%	3C. 24%	4C. 29%	5C. 11%
1D. 31%	2D. 45%	3D. 33%	4D. 41%	5D. 19%
1E. 8%	2E. 10%	3E. 22%	4E. 14%	5E. 53%

SET 10

Olympiad 1	Olympiad 2	Olympiad 3	Olympiad 4	Olympiad 5
1A. 73%	2A. 63%	3A. 74%	4A. 64%	5A. 58%
1B. 80%	2B. 52%	3B. 66%	4B. 39%	5B. 72%
1C. 57%	2C. 29%	3C. 39%	4C. 48%	5C. 34%
1D. 24%	2D. 36%	3D. 22%	4D. 17%	5D. 23%
1E. 5%	2E. 49%	3E. 14%	4E. 28%	5E. 63%

% CORRECT: DIVISION M

SET 11

Olympiad 1	Olympiad 2	Olympiad 3	Olympiad 4	Olympiad 5
1A. 49%	2A. 29%	3A. 49%	4A. 32%	5A. 19%
1B. 24%	2B. 14%	3B. 45%	4B. 18%	5B. 15%
1C. 17%	2C. 34%	3C. 18%	4C. 36%	5C. 19%
1D. 5%	2D. 5%	3D. 5%	4D. 14%	5D. 17%
1E. 21%	2E. 7%	3E. 4%	4E. 1%	5E. 1%

SET 12

Olympiad 1	Olympiad 2	Olympiad 3	Olympiad 4	Olympiad 5
1A. 89%	2A. 73%	3A. 17%	4A. 78%	5A. 56%
1B. 45%	2B. 34%	3B. 32%	4B. 29%	5B. 48%
1C. 78%	2C. 37%	3C. 46%	4C. 71%	5C. 24%
1D. 40%	2D. 59%	3D. 11%	4D. 55%	5D. 33%
1E. 17%	2E. 17%	3E. 6%	4E. 13%	5E. 26%

SET 13

Olympiad 1	Olympiad 2	Olympiad 3	Olympiad 4	Olympiad 5
1A. 65%	2A. 47%	3A. 63%	4A. 49%	5A. 56%
1B. 53%	2B. 44%	3B. 29%	4B. 51%	5B. 32%
1C. 40%	2C. 57%	3C. 35%	4C. 55%	5C. 21%
1D. 46%	2D. 23%	3D. 39%	4D. 31%	5D. 12%
1E. 26%	2E. 19%	3E. 18%	4E. 49%	5E. 56%

SET 14

Olympiad 1	Olympiad 2	Olympiad 3	Olympiad 4	Olympiad 5
1A. 43%	2A. 32%	3A. 45%	4A. 23%	5A. 54%
1B. 35%	2B. 9%	3B. 28%	4B. 40%	5B. 48%
1C. 41%	2C. 48%	3C. 48%	4C. 61%	5C. 34%
1D. 48%	2D. 11%	3D. 43%	4D. 6%	5D. 48%
1E. 11%	2E. 18%	3E. 7%	4E. 11%	5E. 13%

% CORRECT: DIVISION M

SET 15

Olympiad 1	Olympiad 2	Olympiad 3	Olympiad 4	Olympiad 5
1A. 63%	2A. 79%	3A. 68%	4A. 47%	5A. 66%
1B. 60%	2B. 39%	3B. 64%	4B. 45%	5B. 53%
1C. 20%	2C. 38%	3C. 52%	4C. 33%	5C. 14%
1D. 40%	2D. 62%	3D. 41%	4D. 19%	5D. 31%
1E. 22%	2E. 70%	3E. 16%	4E. 33%	5E. 51%

SET 16

Olympiad 1	Olympiad 2	Olympiad 3	Olympiad 4	Olympiad 5
1A. 78%	2A. 45%	3A. 83%	4A. 74%	5A. 82%
1B. 50%	2B. 75%	3B. 79%	4B. 75%	5B. 62%
1C. 52%	2C. 51%	3C. 62%	4C. 67%	5C. 22%
1D. 34%	2D. 36%	3D. 47%	4D. 49%	5D. 42%
1E. 44%	2E. 30%	3E. 37%	4E. 25%	5E. 40%

SET 17

Olympiad 1	Olympiad 2	Olympiad 3	Olympiad 4	Olympiad 5
1A. 71%	2A. 60%	3A. 62%	4A. 52%	5A. 44%
1B. 60%	2B. 60%	3B. 22%	4B. 35%	5B. 49%
1C. 73%	2C. 59%	3C. 26%	4C. 51%	5C. 32%
1D. 24%	2D. 28%	3D. 18%	4D. 28%	5D. 9%
1E. 25%	2E. 16%	3E. 21%	4E. 40%	5E. 25%

SOLUTIONS, STRATEGIES, & FOLLOW-UPS

DIVISION E
SETS 1-10

1A. **METHOD 1:** *Strategy: Rearrange numbers for simplicity.*

$(20 \times 24 \times 28 \times 32) \div (10 \times 12 \times 14 \times 16)$

$$= \frac{20 \times 24 \times 28 \times 32}{10 \times 12 \times 14 \times 16}$$

$$= \frac{20}{10} \times \frac{24}{12} \times \frac{28}{14} \times \frac{32}{16}$$

$$= 2 \times 2 \times 2 \times 2$$

The result of the division is 16.

METHOD 2: *Strategy: Change division into multiplication.*

Dividing by $(10 \times 12 \times 14 \times 16)$ is equivalent to multiplying by the product of their reciprocals, $(\frac{1}{10} \times \frac{1}{12} \times \frac{1}{14} \times \frac{1}{16})$. The expression now becomes $(20 \times 24 \times 28 \times 32) \times (\frac{1}{10} \times \frac{1}{12} \times \frac{1}{14} \times \frac{1}{16})$.

Then using the commutative property to pair the related numbers, we obtain $(20 \times \frac{1}{10}) \times (24 \times \frac{1}{12}) \times (28 \times \frac{1}{14}) \times (32 \times \frac{1}{16})$. Simplifying, the result is $2 \times 2 \times 2 \times 2 = 16$.

1B. **METHOD 1:** *Strategy: Add the appropriate multiple of 8 to 5.*

To reach the twenty-fifth term, Roni must add 24 eights (or 192) to 5. **Thus, Roni's twenty-fifth number is 197.**

METHOD 2: *Strategy: Compare each term to the corresponding multiple of 8.*

If we add 3 to each term, we just have the 8 times table:

Term #	1	2	3	4	5	...	25
Roni's Sequence	5	13	21	29	37	...	?
Multiple of 8	8	16	24	32	40	...	200

$200 - 3 = 197$

FOLLOW-UPS: (1) What is the 1000th number in this sequence? [7997] (2) What is the sum of the first 20 numbers in this sequence? [1620]

1C. **METHOD 1:** *Strategy: Make a tree diagram.*

The 20 segments above represent 20 chords joining any one of the 5 points to each of the other points. But each chord is actually listed twice: for example, $\overline{DE}$ and $\overline{ED}$ are the same chord. Therefore, divide 20 by 2. **We can draw 10 different chords.**

METHOD 2:   *Strategy: Make an organized list.*

After noting that $\overline{DE}$ and $\overline{ED}$ are the same chord, list all chords.
Be sure to list them in order. We can draw 10 different chords.

AB	BC	CD	DE
AC	BD	CE	
AD	BE		
AE			

METHOD 3:   *Strategy: Use combinatorics.*

From each of the 5 points you can draw 4 chords. Since this counts each chord twice, we can draw $5 \times 4 \div 2 = 10$ different chords.

FOLLOW-UP: How many chords would there be if the circle shows 6 points? 7 points? 10 points? Is there a pattern? [15, 21, 45; triangular numbers]

1D. **METHOD 1:**   *Strategy: Adding a number to itself is equivalent to doubling it.*

$$\frac{A + A}{A \times A} = \frac{2 \times A}{A \times A} = \frac{2}{A} \times \frac{A}{A} = \frac{2}{A} \times 1 = \frac{2}{A}$$

Since $\frac{A + A}{A \times A} = \frac{1}{3}$, then $\frac{2}{A} = \frac{1}{3}$. **The value of A is 6.**

METHOD 2:   *Strategy: Make a table.*

A	1	2	3	4	5	6
$A + A$	2	4	6	8	10	12
$A \times A$	1	4	9	16	25	36
$\frac{A + A}{A \times A}$	2	1	$\frac{2}{3}$	$\frac{1}{2}$	$\frac{2}{5}$	$\frac{1}{3}$

The value of A is 6.

1E. **METHOD 1:**   *Strategy: Make a diagram.*

Before	
12	12
x	x
Ben	Jerry

After	
	12
	12
x	x
Ben	Jerry

   *Before: Since Ben gives 12 cards to Jerry, let each start with $x + 12$ cards.
   *After: Since Jerry ends with twice as many cards as Ben, $x = 12 + 12 = 24$.
Thus, Ben starts with 24 + 12 or 36, cards.

METHOD 2:   *Strategy: Work backwards.*

Jerry ends with twice as many cards as Ben. In order to have started with equal amounts, Ben must have given Jerry half the difference in their amounts (see set 1, problem 2A). Since that half is 12, the full difference must be 24. Therefore, Jerry ends with 24 more cards than Ben. This is twice as many cards, so Ben ends with 24 cards. Thus, Jerry ends with 48, and each starts with 36 cards.

SOLUTIONS: DIVISION E

METHOD 3:   *Strategy: Make a table.*
Jerry ends with at least 12 cards and has 2 times as many as Ben, so try numbers beginning with 12 ("After": Jerry). Half of 12 is 6 ("After": Ben). Then work backwards: transfer 12 cards from Jerry to Ben: the results are 18 and 0. Complete the table by increasing the amount for Ben (7, 8, 9, ...) and computing the other three lines until both have the same amount in the "Before" boxes.

AFTER	Ben	6	7	8	9	…	24
	Jerry	12	14	16	18	…	48
BEFORE	Ben	18	19	20	21	…	36
	Jerry	0	2	4	6	…	36

VARIATION: Looking at the "Before" section of the table, you might notice that Jerry's amount increases by 2 for every 1 that Ben's amount increases; that is, Jerry cuts into Ben's "lead" by 1 card per column. Since Ben starts with 18, it takes 18 increases for Jerry to "catch" Ben. Both would then be at 36.

2A. **METHOD 1:**   *Strategy: Let them trade with an imaginary third person.*
At first Marty has 6 pogs more than Jen does. If he gave the 10 pogs to a third person, he then would have 4 pogs fewer than she would. In other words, she would then have 4 pogs more than he. But since Jen receives those 10 pogs, **Jen actually has 14 pogs more than Marty does.**

METHOD 2:   *Strategy: First make their amounts equal.*
After Marty gives Jen 3 pogs, they will have equal amounts, half the total number of pogs each. When Marty gives Jen the other 7 pogs, she will have 7 more than half and he will have 7 less than half, a difference of 14 pogs.

METHOD 3:   *Strategy: Try different numbers in a table to see how the problem works.*

BEFORE			AFTER		
Marty	Jen	Diff.	Marty	Jen	Diff.
10	4	6	0	14	14
11	5	6	1	15	14
12	6	6	2	16	14
13	7	6	3	17	14

METHOD 4:   *Strategy: Make a diagram.*
Before: Since Marty gives Jen 10 pogs, represent his starting amount as $N + 10$. Since he has 6 more than she at the start, represent her starting amount as $N+4$.
After: Transfer the 10-box from Marty's stack to Jen's. Jen now has 14 pogs more than Marty.

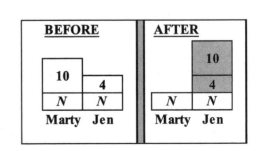

SOLUTIONS: DIVISION E

2B. **METHOD 1:** _Strategy: Draw the figure._
From the diagram, **we see that 6 cubes fit into the box.**

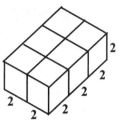

METHOD 2: _Strategy: Compare the volumes of a cube and the box._
Volume of one cube $= 2 \times 2 \times 2 = 8$ cu cm
Volume of the box $= 2 \times 4 \times 6 = 48$ cu cm
Therefore, Michele can fit $48 \div 8 = 6$ cubes into the box.

FOLLOW-UPS: (1) How many faces touch the box? [22] (2) How many cubes would fit if each dimension of the box were multiplied by 2? 3? 4? 10? [48; 162; 384; 6000]

2C. _Strategy: Starting with B, consider every possible value for each unknown number._
$D = 9$ and either $B = 3$ or $B = 7$. If $B = 7$, at least one partial product would have three digits. Neither does, so $B = 3$. Of the 7 combinations of A and C which produce two-digit partial products, only 1 and 2 (in either order) satisfy the condition that the product is a minimum. Then $E = 6$, $F = 3$, and **the least product is 299.**

$$
\begin{array}{r}
A\,B \\
\times\; C\,B \\
\hline
\text{First partial product} \rightarrow \quad E\;9 \\
\text{Second partial product} \rightarrow \quad G\;F \\
\text{Product} \rightarrow \quad \square\,\square\,D
\end{array}
$$

FOLLOW-UP: Look up cryptarithms in the contest problem index on page vii.

2D. _Strategy: Reason from definitions._
A remainder of 4 means that any answer must be a divisor of 18; namely 1, 2, 3, 6, 9 or 18. It also means that the divisor must be greater than 4. **Thus, the "eligible" divisors are 6, 9, or 18.**

FOLLOW-UP: More remainder problems are listed in the contest problem index on page ix.

2E. **METHOD 1A:** _Strategy: Consider the difference between the maximum and actual incomes._
A group of all adults would pay $84, which is $20 too much. Since each child pays $4 less than each adult does, divide $20 by $4. **Thus, there are 5 children in the group.**

METHOD 1B: _Strategy: Consider the difference between the minimum and actual incomes._
A group of all children would pay $36, which is $28 too little. Since each adult pays $4 more than each child does, divide $28 by $4 to get 7 adults in the group. There are 5 children in the group.

METHOD 2: *Strategy: Set up a table.*

The first column shows the number of children and the third column shows the corresponding number of adults. The second and fourth columns show the amounts of money paid in each case. The last column shows the total amount of money paid.

Children		Adults			TOTAL
0	... $0	12	...	$84	$84
1	... 3	11	...	77	80
2	... 6	10	...	70	76
3	... 9	9	...	63	72
4	... 12	8	...	56	68
5	... $15	7	...	$49	$64

3A. *Strategy: Consider the time every 12 hours from now.*

At the end of 96 hours (8 groups of 12 hours each), the clock will still show 10:45, with 4 hours left over from the original 100 hours. **4 hours after 10:45 is 2:45.**

FOLLOW-UP: If today is Tuesday, what day of the week will it be 1000 days from now? [Monday]

3B. **METHOD 1:**   *Strategy: Count vertical stacks and add.*

4 columns of 1 each   = 4
2 columns of 2 each   = 4
2 columns of 3 each   = 6
1 column of 4   = 4
1 column of 5   = 5
The tower contains 23 cubes.

METHOD 2:   *Strategy: Count horizontally from the top and add.*

Top layer: 1
Second layer: 2
Third layer: 4
Fourth layer: 6
Bottom layer: 10
The tower contains 23 cubes.

3C. **METHOD 1:** *Strategy: Consider the number of games the Panthers have lost.*

The team played 9 games, winning 2 and losing 7. To win exactly half of its games, it must have won a total of 7 games. **The number N represents 5 , the number of remaining games.**

METHOD 2: *Strategy: Make a table.*

In the table below, N represents 0, 1, 2, and so on. Fractions are formed comparing the resulting numbers of games won to games played. This is continued until a fraction has a value of $\frac{1}{2}$.

N	0	1	2	3	4	5
$\dfrac{\text{Games won}}{\text{Games played}}$	$\dfrac{2}{9}$	$\dfrac{3}{10}$	$\dfrac{4}{11}$	$\dfrac{5}{12}$	$\dfrac{6}{13}$	$\dfrac{7}{14}=\dfrac{1}{2}$

N represents 5.

3D. **METHOD 1:** *Strategy: Consider the common multiples of 4 and 5.*

If we had 3 students more, there would be no students left over for groups of 4 or for groups of 5. The positive multiples of 4 and 5 are 20, 40, 60, etc. If we now subtract the three students we added earlier, our possible answers will be 17, 37, 57, etc. **Then, the least number of students the class could have is 17.**

METHOD 2: *Strategy: Use the given conditions to consider only possible numbers.*

Since, after division by 5, the remainder is 2, the units digit is either 2 or 7. Since after division by 4, the remainder is 1, the units digit is odd, making the units digit 7. Divide 7, 17, 27, ... by 4. There are at least 17 students in the class.

3E. **METHOD 1:** *Strategy: Represent visually.*

Let each $\square = \frac{1}{3}$ of the number, so that 3 boxes = the number. Then:

$\square + 16 = \square\,\square\,\square \quad \square\,\square\,\square \quad \square\,\square\,\square$

$16 = \square\,\square\,\square \quad \square\,\square\,\square \quad \square\,\square$

Since 8 boxes represent 16, one box represents 2.

Since 3 boxes represent the number itself, **the number is 6.**

METHOD 2: *Strategy: Reasoning.*

If we add $\frac{1}{3}$ of a number to 16, we get the number three times. Therefore, 16 alone must equal $2\frac{2}{3}$ times the number. This is $\frac{8}{3}$ of the number. If $\frac{8}{3}$ of the number is 16, then $\frac{1}{3}$ of the number is 2 and the number is 6.

4A.   **METHOD 1:**   _Strategy: Use cancellation in fractions._

$\frac{1}{2}$ of $\frac{2}{3}$ of $\frac{3}{4}$ of $\frac{4}{5}$ of 100 means $\frac{1}{2} \times \frac{2}{3} \times \frac{3}{4} \times \frac{4}{5} \times 100$.

Then cancel the 2s, the 3s and the 4s. Each cancellation is indicated by a different figure.

$\frac{1}{2} \times \frac{2}{3} \times \frac{3}{4} \times \frac{4}{5} \times 100 = \frac{1}{5} \times 100 = 20$. **The value of _N_ is 20.**

METHOD 2:   _Strategy: Work right to left._

$\frac{4}{5}$ of $100 = 80$

$\frac{3}{4}$ of $80 = 60$

$\frac{2}{3}$ of $60 = 40$

$\frac{1}{2}$ of $40 = 20$

METHOD 3:   _Strategy: Work left to right._

$\frac{1}{2} \times \frac{2}{3} = \frac{1}{3}$

$\frac{1}{3} \times \frac{3}{4} = \frac{1}{4}$

$\frac{1}{4} \times \frac{4}{5} = \frac{1}{5}$

$\frac{1}{5} \times 100 = 20$

4B.   **METHOD 1:**   _Strategy: Add areas._

Each small square has an area of 1. Separate _EFBA_ into rectangle I and right triangles II and III, as shown. The area of rectangle I is 6; triangle II is $\frac{1}{2}$ of rectangle _AGED_, so its area is 1.5. The same is true of triangle III. Thus, $6 + 1.5 + 1.5 = 9$. **The area of trapezoid _EFBA_ is 9 sq units.**

METHOD 2:   _Strategy: Subtract areas._

Rectangle _ABCD_ has an area of 12. Triangle _X_ is $\frac{1}{2}$ of rectangle _AGED_ so its area is 1.5. The same is true of triangle _Y_. Thus, the area of trapezoid _EFBA_ is $12 - 1.5 - 1.5 = 9$.

METHOD 3:   _Strategy: Combine like shapes._

Modify either method by moving one of the triangles and joining it to the other as shown.

FOLLOW-UP: Rectangle ABCD at right contains 16 congruent squares. The area of ABCD is 144 sq cm. Find the area of triangle PDC. [72 sq cm.]

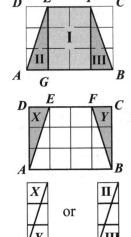

4C. **METHOD 1:** *Strategy: Make a list.*

Each factor of a given number has a "cofactor" such that the product of the factor and its cofactor is the given number. To catch all the larger factors, pair each factor of 36 with its cofactor in order:

$$1 \times 36$$
$$2 \times 18$$
$$3 \times 12$$
$$4 \times 9$$
$$6 \times 6 \qquad \textbf{36 has 9 different factors.}$$

METHOD 2: *Strategy: Express 36 as a product of prime numbers.*

If p is a prime number, then p^2 has 3 factors: 1, p and p^2.

Notice that $36 = 2 \times 2 \times 3 \times 3 = 2^2 \times 3^2$. The factors of 2^2 are 1, 2, and 4. The factors of 3^2 are 1, 3, and 9. We get every factor by multiplying each of 1, 2, and 4 by each of 1, 3, and 9. Thus, 3 factors paired with each of 3 other factors gives us 9 different products.

$$1 \times 1, \quad 1 \times 3, \quad 1 \times 9;$$
$$2 \times 1, \quad 2 \times 3, \quad 2 \times 9;$$
$$4 \times 1, \quad 4 \times 3, \quad 4 \times 9. \qquad \text{The number has 36 different factors.}$$

FOLLOW-UP: What is the largest odd factor of 336? [21]

4D. **METHOD 1:** *Strategy: Work only with integral values for both time and distance.*

$1\frac{1}{2}$ minutes	⇨	1 mile	(Double both quantities)
3 minutes	⇨	2 miles	(Multiply both quantities by 20)
60 minutes	⇨	40 miles	

In one hour the car will travel 40 miles.

METHOD 2: *Strategy: Find the number of 1.5 minute intervals in 1 hour.*

The number of 1.5 minute intervals in 60 minutes is $60 \div 1.5$ (or $60 \div 1\frac{1}{2}$) = 40. Since the car travels 1 mile in each of those 40 intervals, then the car travels 40 miles in one hour.

4E. **METHOD 1:** *Strategy: Combine purchases to get the same number of pens and pencils.*

3 pens and 2 pencils sell for 47¢ while 2 pens and 3 pencils sell for 38¢. Combining the two purchases, 5 pens and 5 pencils sell for 85¢. Dividing by 5, **the cost of a set of one pen and one pencil is 17¢.**

METHOD 2: *Strategy: Compare the costs of 1 pen and 1 pencil.*

Compare the two purchases: 1 pen sells for 9¢ more than 1 pencil. Thus in the first purchase, 3 pens sell for 27¢ more than 3 pencils, so that 5 pencils sell for 27¢ less than 3 pens and 2 pencils. That is, 5 pencils sell for $47 - 27 = 20$¢, and 1 pencil sells for $20 \div 5 = 4$¢.

Return to the first purchase: since a pencil sells for 4¢, a pen must sell for 9¢ more, or 13¢. Therefore, the set sells for $4 + 13 = 17$¢. Check that 4¢ and 13¢ satisfy the two purchases.

METHOD 3: *Strategy: Make a table.*
Compare the two purchases to see that 1 pen sells for 9¢ more than 1 pencil. Then make a table using this fact:

COSTS			
1 pen	1 pencil	2 pens + 3 pencils	3 pens + 2 pencils
10¢	1¢	23¢	32¢
11	2	28	37
12	3	33	42
13	4	38	47

Don't forget to add 4¢ + 13¢ = 17¢ in order to answer the question.

FOLLOW-UPS: (1) There are 50 horses and riders. They have a total of 128 legs. How many horses are there? [14] (2) Toni has 35 dimes and quarters with a total value of $7.10. How many more quarters than dimes does she have? [13]

5A. **METHOD 1:** *Strategy: Multiples of 10 are also multiples of 5.*
The number of terminal zeroes equals the number of factor pairs of 5 and 2. 10 appears as a factor 5 times to yield 5 terminal zeroes plus those created by multiplying $3 \times 4 \times 5 \times 6 \times 7$. The product of the only 5 and any even number will end in zero, adding on one more terminal zero. Since there is only one five in $3 \times 4 \times 5 \times 6 \times 7$, **the product has 6 terminal zeroes.**

METHOD 2: *Strategy: Multiply out.*
$30 \times 40 \times 50 \times 60 \times 70 = 252,000,000$. This product has 6 terminal zeroes.

FOLLOW-UP: What prime numbers are factors of the product of all the counting numbers which are less than 20? [2, 3, 5, 7, 11, 13, 17, 19]

5B. *Strategy: Separate the squares.*
The diagram is the result of combining these two figures. Each has 4 small squares and 1 large square for a total of 5 squares. $5 + 5 = 10$. **There is a total of 10 squares of all sizes.**

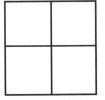 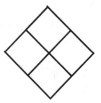

FOLLOW-UPS: How many squares of all sizes can be drawn using the lines of a checkerboard (an 8-by-8 square)? [204] How many rectangles? [1296]

5C. _Strategy: Consider the larger denomination (quarters) first._

In order to make $1.95, there must be fewer than 8 quarters. There cannot be 2, 4, or 6 quarters because the balances of $1.45, $.95, and $.45 cannot be made using only dimes. So 1, 3, 5 and 7 quarters are possible, calling for 17, 12, 7, and 2 dimes, respectively. **Megan can have 4 different values of _A_.**

5D. _Strategy: Draw the clock and construct a table._

Number	1	2	3	4	5
Opposite	6	7	8	9	10
Sum	7	9	11	13	15

3 and 8 have a sum of 11.

FOLLOW-UP: Suppose the hours on another alien clock were marked from 7 to 52, inclusive. What would be opposite 32? [55]

5E. **METHOD 1:**   _Strategy: Make the approach visual._

STORE A	STORE B	STORE C: $12	At the start

Spent	Left Over	Left Over

$\Rightarrow$

	Spent
	Left Over
	Left Over

$\Rightarrow$

	$3	$3
	$3	$3

$\Rightarrow$

$3	$3	$3
$3	$3	$3
$3	$3	$3

Gil had $27 at the start.

METHOD 2:   _Strategy: Work backwards._

Gil has $12 left from store B. This is $1 - \frac{1}{3} = \frac{2}{3}$ of the money left from store A. Thus he left store A and entered store B with $12 \div \frac{2}{3} = $18. Similarly, the $18 he had when he left store A was $\frac{2}{3}$ of the money he started with. He entered store A with $18 \div \frac{2}{3} = $27. Gil started with $27.

SET 2	◆ ◆ ◆	Olympiad 1

1A. _Strategy: Group consecutive pairs of numbers._

$$9 + 91 = 100$$
$$18 + 82 = 100$$
$$27 + 73 = 100$$
$$36 + 64 = 100$$
$$\underline{45 + 55 = 100}$$
Sum = 500

SOLUTIONS: DIVISION E

1B. *Strategy: Examine the stairway from each of the three dimensions.*

  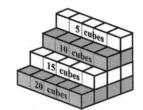

The end "slice" on the right has 10 cubes. 5 such slices = **50 cubes.**

Add the four stacks: $5 + 10 + 15 + 20 = 50.$

Add the four layers: $20 + 15 + 10 + 5 = 50.$

1C. **METHOD 1:** *Strategy: Treat each purchase as a single entity.*
ToyWorld: At 4 crayons for 25¢, 5 purchases buy 20 for $1.25.
GameLand: At 5 crayons for 30¢, 4 purchases buy 20 for $1.20.
Therefore, **GameLand charges 5¢ less for 20 crayons**.

METHOD 2: *Strategy: Compare unit costs.*
ToyWorld: 1 crayon costs $25 \div 4 = 6\frac{1}{4}$¢ each.
GameLand: 1 crayon costs $30 \div 5 = 6$¢ each.
Thus, for each crayon, GameLand charges $\frac{1}{4}$¢ less.
For 20 crayons, GameLand charges $20 \times \frac{1}{4}$¢ = 5¢ less than ToyWorld.

Follow-Up: *A baseball player gets 151 hits in his first 453 times at bat. In his next game he gets 2 hits in 5 at bats. Does his batting average rise or fall? Explain.* [Rise; his .400 batting average (2 out of 50) in that last game is higher than his previous .333 batting average (151 out of 453), and therefore, raises it.]

1D. **METHOD 1:** *Strategy: Determine how many students were counted twice.*
Since 3 students do not like either flavor, exactly 23 like one or both flavors. Since 15 like Vanilla and 16 like Chocolate, a total of 31 have been counted. So $31 - 23 = 8$ students must have been counted twice, once for each flavor. Thus, **8 students like both flavors.**

METHOD 2: *Strategy: Use a Venn Diagram.*
Each region represents the members of that category. Region V represents those who like Vanilla only; C = Chocolate only; B = both flavors; N = neither flavor.

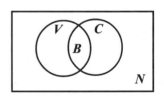

Region N has 3 students, so $V + C + B$ has $26 - 3 = 23$, students. But circle $C + B$ is given as 16 (all the Chocolate Lovers), so V alone must have $23 - 16 = 7$ students (Vanilla only). Since $V + B$ is given as 15 (all the Vanilla lovers), B must have $15 - 7 = 8$ students. 8 students like both flavors.

METHOD 3: *Strategy: Use a table.*

Let line 1 represent those who like both, and try the values 0, 1, 2, 3, and so on. Obtain the values in line 2 by subtracting each entry in row 1 from 15. Obtain the values in line 3 by subtracting each entry in row 1 from 16. Add lines 1, 2, and 3 to get row 4; this is the

Both	0	1	2	...	?
Vanilla Only	15	14	13	...	
Chocolate Only	16	15	14	...	
Total no. of people	31	30	29	...	23

total number of people who like at least one flavor. Note that the sum of rows 4 and 1 is always 31. When the entry in row 4 is 26 – 3 = 23, then the entry in row 1 is 31 – 23 = 8.

1E. METHOD 1: *Strategy: "Spread" Todd's 12 extra pounds among the group.*

Todd weighs 12 pounds more than the average of the original group. This raises the average by 2 pounds, from 100 to 102. It is convenient to pretend that the 12 pounds could be distributed equally among all the children, 2 lbs each. Divide the 12 lb total by the 2 lb per child to get 6 children in the new group, including Todd. Thus, **there were 5 children in the original group.**

METHOD 2: *Strategy: Compare Todd's weight to the old average weight.*

Since Todd's weight exceeds the new average weight by 10 pounds, the 10 pounds will raise the original group's average of 100 to 102. There must be 10 ÷ 2 = 5 children in the original group.

FOLLOW-UP: Exploration: Assume, before Todd joins, that each member of the group weighs 100 pounds and that there are 1, 2, 3, ... members. Investigate the effect of Todd's 112 pounds on the average of the group as it gets larger.

SET 2   Olympiad 2

2A. *Strategy: First calculate N.*

Ana: If $N \div 8 = .25$, then $N = 8 \times .25 = 2$.

Barney: $N \times 8 = 2 \times 8 = 16$.

Barney's answer was 16.

2B. METHOD 1A: *Strategy: Use a visual approach.*

Let box J represent the number of computer games Jay has. Then Sherry's collection is represented by a stack of five boxes, Nairan's collection by a stack of three boxes, and Sherry has two more boxes than Nairan. But since we are told that Sherry has 16 more games than Nairan, the two extra boxes represent 16 games, and each box represents 8 games. **Jay has 8 computer games.**

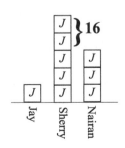

SOLUTIONS: DIVISION E

METHOD 1B: *Strategy: Use algebra.*
Suppose Jay has *J* computer games. Then Sherry has 5*J* games and Nairan has 3*J* games. It is given that 2*J* is 16. Then Jay has 8 computer games.

METHOD 2:   *Strategy: Make a table.*
Let row 1 represent all choices for Jay: 1, 2, 3, and so on. Then rows 2 and 3 contain the possibilities for Sherry and Nairan. Subtract row 3 from row 2 to get row 4. Note that the entries in row 4 are twice the corresponding entries in row 1. The difference is given as 16. Thus, Jay has 8 games.

Jay	1	2	3	...	?
Sherry	5	10	15	...	
Nairan	3	6	9	...	
Difference	2	4	6	...	16

FOLLOW-UP: If a number is added to itself and then increased by 60, the result is equal to seven times the original number. What is 2 less than 3 times the number? [34]

2C. *Strategy: Start with 5 and 1.*
To get a difference of 5, *B* must be 6.
Then *C*, *D*, and *A* are 2, 3, and 4 in some order.
D is one less than *C*, so *C* is 3 or 4.
Suppose *C* = 3. Then *D* = 2 and *A* = 3. But *A* cannot equal *C*.
Thus, *C* = 4. This makes *D* = 3 and **A = 2**. All 6 numbers are used, each once.

FOLLOW-UP: Exploration: Research magic triangles.

2D. *Strategy: Combine the areas of the nine shaded regions in the interior.*
Each shaded triangle marked *X* is half of a small square, so the four *X*-triangles are equal in area to 2 full squares. Combined with the 5 shaded full squares, the total shaded area is equal to 7 squares. But the area of the shaded region is given as 14 square centimeters. Thus each small square has an area of 2 square cm and **the area of *ABCD* is** $9 \times 2 = $ **18 sq cm.**

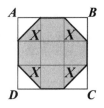

FOLLOW-UPS: (1) Compute the length of one side of a small square to the nearest tenth. [1.4] *(2) Discussion: How can the area of a square be something other than a square number?*

2E. **METHOD 1:**   *Strategy: Use the reasoning of combinations.*
Suppose there are 10 people. Each shakes hands with 9 other people. But every handshake is counted twice because if *B* is on *A*'s list of handshakes, then *A* is on *B*'s list. To get the total number of handshakes, multiply 10 by 9 and divide by 2. Applying this to the given problem, double 15. The 30 we get must equal the product of two consecutive numbers, which are 6 and 5. **6 people are in the room.**

METHOD 2
All of the following methods are actually the same method mathematically. The differences are in how one sees the situation.

SOLUTIONS: DIVISION E

METHOD 2A: _Strategy: Act it out with smaller numbers._

Two people alone in the room shake hands. Total = 1 handshake.

A third person enters and shakes with both: Add 2 more: Total = 3 handshakes.
A fourth person joins them and shakes with each: Add 3 more: Total = 6 handshakes.
A fifth person joins them and shakes with each: Add 4 more: Total = 10 handshakes.
A sixth person joins them and shakes with each: Add 5 more: Total = 15 handshakes.

For 15 handshakes, there are 6 people in the room.

METHOD 2B: _Strategy: Create an organized visual table and keep count._

Represent the people by A, B, C, and so on.

Two people shake:	_BA_					Total so far = 1.
The third joins in:	_CA_	_CB_				Total so far = 3.
The fourth joins in:	_DA_	_DB_	_DC_			Total so far = 6.
The fifth joins in:	_EA_	_EB_	_EC_	_ED_		Total so far = 10.
The sixth joins in:	_FA_	_FB_	_FC_	_FD_	_FE_	Total so far = 15.

There are 6 people in the room.

(Note: The sums 1, 3, 6, 10, 15, and so on are called Triangular Numbers and present much opportunity for enrichment.)

METHOD 2C: _Strategy: Relate to a problem you've seen before._

This problem is mathematically identical to problem 1C of Set 1. Replace the people by points on a circle and the handshakes by chords of the circle.

2 points determine 1 chord.
3 points determine 3 chords.
4 points determine 6 chords.
5 points determine 10 chords.
6 points determine 15 chords.

Thus, there are 6 people in the room.

SET 2	Olympiad 3

3A. _Strategy: Divide both sides by common factors to simplify the question._

If we divide two equal quantities by the same number, our results will be equal.

Original statement:	$1 \times 2 \times 3 \times 4 \times 5 \times 6 \times 7 = 8 \times 9 \times 10 \times N.$
Divide both sides by 2:	$1 \times 3 \times 4 \times 5 \times 6 \times 7 = 4 \times 9 \times 10 \times N.$
Divide both sides by 3:	$1 \times 4 \times 5 \times 6 \times 7 = 4 \times 3 \times 10 \times N.$
Divide both sides by 4:	$1 \times 5 \times 6 \times 7 = 1 \times 3 \times 10 \times N.$
Divide both sides by 5:	$1 \times 6 \times 7 = 3 \times 2 \times N.$
Divide both sides by 3:	$2 \times 7 = 1 \times 2 \times N.$
Divide both sides by 2:	$1 \times 7 = 1 \times N.$

Thus, $7 = N$, therefore, **the value of N is 7.**

(Note: Divide out several factors at once to speed up the process. This is related to cancellation).

3B. **METHOD 1:** _Strategy_: _Examine common multiples of 3 and 5._
Any multiple of 3 and 5 is also a multiple of 15, their least common multiple. Look at all multiples of 15 between 121 and 149. Since $15 \times 8 = 120$ and $15 \times 10 = 150$ are not between 121 and 149, $15 \times 9 = 135$ is the only possibility. **The number is 135.**

METHOD 2: _Strategy_: _Construct a table of multiples._
Compare all multiples of 3 with all multiples of 5, in the interval between 121 and 149:
  3: 123, 126, 129, 132, 135, 138, 141, 144, 147
  5: 125, 130, 135, 140, 145
Only 135 appears on each list.

FOLLOW-UP: _What is the least counting number that is a multiple of both 2 and 7, and also a square number?_ [196]

3C. **METHOD 1:** _Strategy_: _Count up, using round numbers to keep it simple._
From 7:45 am to 8:00 am: 0 hrs. 15 min.
From 8:00 am to 12 noon: 4 hrs. 0 min.
From 12: noon to 4:00 pm: 4 hrs. 0 min.
From 4:00 pm to 4:05 pm: 0 hrs. 5 min.
She is gone for 8 hrs. 20 min.

METHOD 2: _Strategy_: _Subtract, using regrouping techniques._

4: 05 pm		16:05		"15:65"
− 7: 45 am	→	− 7:45	→	− 7:45
				8:20

3D. _Strategy_: _Use the perimeter of the square to find the perimeter of the rectangle._
For the square, the area is 36. Then the length of one side is 6 and the perimeter is 24. For the rectangle, a perimeter of 24 implies that the sum of the length and width (the semiperimeter) is 12. The length is twice the width, so the length is 8 and the width is 4. Thus the area is $8 \times 4 = 32$. **The number of square cm in the area of the rectangle is 32.**

3E. _(Note: So far, 12 different methods to solve this problem have been found. Two of them are offered here. How many others can you find?)_
METHOD 1: _Strategy_: _Use "Gaussian Addition"._
The twentieth even counting number is 40, so the twentieth odd counting number is 39. The first twenty odd numbers are 1, 3, 5, …, 39. Then combine $1 + 3 + 5 + \ldots + 35 + 37 + 39$ in pairs, from the outside inward: $(1 + 39) + (3 + 37) + (5 + 35) + \ldots + (19 + 21)$. The result is 10 sums, each equal to 40. **The first twenty rows contain 400 stars.**

METHOD 2:   *Strategy: Work from simpler cases.*
Set up a table and look for a pattern.

Number of Rows	1	2	3	4	5	...	20
Total Number of Stars	1	4	9	16	25	...	?

Note that $1 \times 1 = 1,$
                    $2 \times 2 = 4,$
                    $3 \times 3 = 9,$
                    $4 \times 4 = 16,$ and so on.
The total number of stars by the end of each row is the square of the number of rows to that point. By this pattern, the first 20 rows will have a total of $20 \times 20 = 400$ stars. 400 stars are contained in the first 20 rows.

4A.   *Strategy: Use the definition of prime number.*
By definition, a prime number is any counting number that has exactly two factors: the number one and the number itself. Thus, 1 is not a prime number. The prime numbers less than 15 are: 2, 3, 5, 7, 11, and 13. **6 counting numbers less than 15 have exactly two factors.**

FOLLOW-UPS: (1) List all counting numbers less than 50 that have exactly three factors. [1, 4, 9, 25, 49] (2) Describe the complete set of counting numbers that have exactly three factors. [They are the squares of prime numbers.]

4B.   METHOD 1:   *Strategy: Compare the widths in the drawing and the actual room.*
6 cm represents 15 feet, so every 2 cm represents 5 feet. In the drawing the length is given as 10 cm, which is 5×2 cm. This represents 5×5 feet = 25 feet. Therefore, **the actual perimeter is** $2 \times 25 + 2 \times 15 =$ **80 feet.**

METHOD 2:   *Strategy: Form a proportion and use algebra.*
The perimeter of the scale drawing is 32 cm. To find the actual perimeter, write a proportion and use the theorem that in a proportion, the cross products are equal:

$$\frac{6 \text{ cm}}{15 \text{ feet}} = \frac{32 \text{ cm}}{L \text{ feet}}$$
$$6 \times L = 15 \times 32$$
$$6 \times L = 480$$
$$L = 80$$

The actual perimeter is 80 feet.

METHOD 3:   *Strategy: Compare the length to the width.*

In the scale drawing, the length is $\frac{5}{3}$ times the width. Thus in the actual room the length will also be $\frac{5}{3}$ times the width. $\frac{5}{3} \times 15$ feet $= 25$ feet. Then proceed as in Method 1.

4C. *Strategy: Examine simpler cases.*

2^1 is **2**	2^5 ends in **2**	2^9 ends in **2**	2^{13} ends in **2**
2^2 is **4**	2^6 ends in **4**	2^{10} ends in **4**	2^{14} ends in **4**
2^3 is **8**	2^7 ends in **8**	2^{11} ends in **8**	2^{15} ends in **8**
2^4 is **16**, which ends in **6**	2^8 ends in **6**	2^{12} ends in **6**	2^{16} ends in **6**

The unit's digit repeats in groups of four: 2, 4, 8, 6. So 2^4, 2^8, 2^{12}, 2^{16}, 2^{20}, and every power of 2 whose exponent is a multiple of four has a units digit of 6. Therefore 2^{1996} has a units digit of 6 and 2^{1997} **has a units digit of 2.**

FOLLOW-UPS: Explorations: If each of the following is multiplied out, what is the units digit of the product? 3^{1997}, 4^{1997}, 5^{1997}, 6^{1997}, 7^{1997}, 8^{1997}, 9^{1997}? [3, 4, 5, 6, 7, 8, 9]

4D. **METHOD 1A:**   *Strategy: Suppose both people buy the same number of shirts.*

Suppose Marc makes his purchase 3 times. Then each person buys the same number of shirts and we can compare the amounts spent on caps:

Thus: Marc's 3 purchases: 3 shirts and 15 caps cost $69.
Kathy's 1 purchase: 3 shirts and 2 caps cost $30.

Compare these purchases: Marc buys 13 more caps than Kathy for an extra $39. Thus 1 cap costs $3. Then Kathy bought 2 caps for $6 and had $24 left for 3 shirts. Therefore, **each shirt costs $8.**

METHOD 1B: *Strategy: Suppose both people buy the same number of caps.*

Suppose Kathy makes her purchase 5 times and Marc makes his twice. Then each person buys the same number of caps and we can compare the amounts spent on shirts:

Thus, Kathy's 5 purchases: 15 shirts and 10 caps cost $150.
Marc's 2 purchases: 2 shirts and 10 caps cost $46.

Compare these purchases: Kathy buys 13 more shirts than Marc for an extra $104. Each shirt costs $8.

(Note: The methods above follow the usual alebraic techniques very closely.)

FOLLOW-UPS: Select word problems from any Algebra 1 textbook and solve without algraic tools.

4E. _Strategy_: _Decide what values are possible and then substitute._

To produce an end result of 0, $12 + (\square \times \square)$ must be equal to $(7 \times \square)$. Therefore $(\square \times \square)$ is less than $(7 \times \square)$, and $\square$ is less than 7. Thus $\square$ represents the whole numbers 0, 1, 2, 3, 4, 5, or 6. Replace all three boxes with the same number until two different choices work.

$\square$	$12 + (\square \times \square) - (7 \times \square)$	Computation	Result = zero?
0	$12 + (0 \times 0) - (7 \times 0)$	$12 + \ 0 - \ \ 0$	No
1	$12 + (1 \times 1) - (7 \times 1)$	$12 + \ 1 - \ \ 7$	No
2	$12 + (2 \times 2) - (7 \times 2)$	$12 + \ 4 - 14$	No
3	$12 + (3 \times 3) - (7 \times 3)$	$12 + \ 9 - 21$	YES; $' = 3$
4	$12 + (4 \times 4) - (7 \times 4)$	$12 + 16 - 28$	YES; $' = 4$

The two numbers are 3 and 4.

SET 2 Olympiad 5

5A. _Strategy_: _Eliminate the faces that touch the face that contains T._

Let each letter represent the square face that contains it. T touches both A and H. After folding, both S and E also touch T. That leaves only M. Therefore, **M is on the face opposite the letter T.**

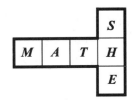

5B. **METHOD 1:** _Strategy_: _Start by comparing other ages to that of Billy._

Trini is 1 month older than Billy and Jason is 3 months older than Billy. Thus, Jason is two months older than Trini. Kimberly is 4 months older than Jason, so **Kimberly is 6 months older than Trini.**

METHOD 2: _Strategy_: _Make a diagram._

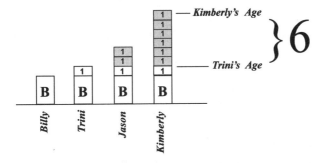

Kimberly is 6 months older than Trini.

METHOD 3:   *Strategy: Make a table and look for a pattern.*
Assume random samples for Billy's age (say, 100 months, 82 months, and 29 months) and compute the differences between Kimberly's and Trini's ages. For example:

Billy	**Trini**	**Jason**	**Kimberly**	**Differences**
100	101	103	107	$107 - 101 = 6$
82	83	85	89	$89 - 83 = 6$
29	30	32	36	$36 - 30 = 6$

Kimberly is 6 months older than Trini.

5C. **METHOD 1:**   *Strategy: Reason from right to left.*
1. From the given choices, only $D = 6$ and $H = 9$ can produce 7. Regrouping occurs.
2. Because of regrouping, C and G must be the same. Since 9 was used in step 1, $C = G = 4$. Regrouping occurs.
3. Because of regrouping, B and F differ by 5. From the remaining choices, only $B = 3$ and $F = 8$ can produce 5. Regrouping occurs.
4. By elimination, $A = 9$ and $E = 5$. Thus $ABCD = 9346$ and $EFGH = 5849$.
The boxes represent 9346.

$$\begin{array}{r} A\ B\ C\ D \\ -\ E\ F\ G\ H \\ \hline 3\ 4\ 9\ 7 \end{array}$$

METHOD 2:   *Strategy: Change subtraction to addition.*
Doing the problem as an addition, right to left, is the clearest, easiest method. It is very similar to Method 1. However, adding left to right (working backwards) illustrates the process of elimination:
1. If $E = 8$ or 9, then A has two digits, which is not allowed. If $E = 4$, then A ends in 7 or 8; neither choice is available. Thus $E = 5$, $A = 9$ and regrouping has occurred.
2. $F \neq 4$ because $F + 4$ does not require regrouping. Then $F = 8$ or 9.
3. Four possibilities exist for $EFGH$: 5849, 5894, 5948, and 5984. Adding 3497 to each in turn produces digits which are not available for 3 of the 4 possibilities.
The boxes represent 9346.

$$\begin{array}{r} 3\ 4\ 9\ 7 \\ +\ E\ F\ G\ H \\ \hline A\ B\ C\ D \end{array}$$

FOLLOW-UPS: Select cryptarithms from pages 115-118 of "Creative Problem Solving in School Mathematics".

5D. **METHOD 1:**   *Strategy: Visualize the situation.*
The perimeter of the pool is $2 \times 30 + 2 \times 20 = 100$ feet. We need $100 \div 2 = 50$ of the 2 by 2 tiles to construct the border of the pool. But this does not account for the four corners, which require 4 additional tiles to complete the walkway.
The fewest number of tiles is 54.

METHOD 2: _Strategy: Compare areas._
The length of the combined pool and walkway is at least 34 feet and the width, at least 24 feet. Thus the combined area is at least $34 \times 24 = 816$ square feet. The area of the pool alone is $30 \times 20 = 600$ square feet. Therefore, the smallest area of the walkway alone is their difference, 216 square feet. Since the area of each tile is 4 square feet, then $216 \div 4 = 54$ tiles are needed to build the walkway.

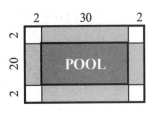

FOLLOW-UP: The area of a rectangle is 225 sq cm. The greatest common factor (GCF) of its length and width is 1. If both dimensions are whole numbers less than 100 cm, what is the perimeter of the rectangle? [68 cm]

5E. **METHOD 1:** _Strategy: Work forwards._
Tyler receives $\frac{1}{2}$ of Alexandra's marbles. Therefore Jan receives $\frac{2}{3}$ of $\frac{1}{2} = \frac{1}{3}$ of Alexandra's marbles and Jerry receives $\frac{3}{4}$ of $\frac{1}{3} = \frac{1}{4}$ of Alexandra's marbles. The least whole number which is divisible by the denominators 2, 3, and 4 is 12.
The fewest number that Alex can start with is 12.

METHOD 2: _Strategy: Work backwards._
The fewest marbles that Jerry can receive is $\frac{3}{4}$ of $4 = 3$. So Jan had 4 marbles. Since 4 is $\frac{2}{3}$ of Tyler's amount, Tyler had 6. Since 6 is $\frac{1}{2}$ of Alexandra's amount, then the fewest number that Alex can start with is 12.

METHOD 3
Strategy: Choose a convenient value to start and then adjust your final figures.
Choose any multiple of all three denominators: suppose Alexandra has, say, 60 marbles. Then Tyler has $\frac{1}{2}$ of $60 = 30$ marbles; Jan has $\frac{2}{3}$ of $30 = 20$ marbles; and Jerry has $\frac{3}{4}$ of $20 = 15$ marbles. Since the greatest common factor of 60, 30, 20, and 15 is 5, we can divide each one's share by 5. Thus, Alexandra starts with 12, Tyler gets 6, Jan gets 4, and Jerry gets 3. The fewest number of marbles that Alex can start with is 12.

FOLLOW-UP: If $\frac{1}{2}$ of $\frac{2}{3}$ of $\frac{3}{4}$ of $\frac{4}{5}$ of a number is 20, what is the number? [100]

SET 3 ◆ ◆ ◆ **Olympiad 1**

1A. _Strategy: Group for easier addition._
Add the numbers left of the equals sign in pairs. Each of the first four pairs has a sum of 100: $82 + 18 = 100$, $83 + 17 = 100$, $84 + 16 = 100$, and $85 + 15 = 100$. So the sum of the four pairs is 400. Since the sum of all the numbers is 500, $71 + N = 500 - 400 = 100$. Therefore **the value of the counting number N is $100 - 71 = 29$.**

SOLUTIONS: DIVISION E

1B. *Strategy: Keep filling in boxes until A is determined.*
Since the sum of all the numbers from 1 to 9 is 45, the sum of the three
numbers along any one line is 15. The positions of 3 and 7 places 5 in the
center. Then the positions of 1 and 5 places 9 as shown. This leaves 2, 4,
6, or 8 as the corner numbers.

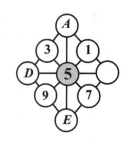

Let D and E represent the numbers in the circles as shown. If $D = 6$ or 8,
then $D + 9 + E$ is more than 15. If $D = 2$, then A is 10, which is not a
choice. Thus, $D = 4$, and **the number in the circle marked A is 8.**

*FOLLOW-UP: Do research on Magic Squares and on Magic Triangles. Construct a Magic Triangle all of
whose entries are the odd integers from 13 through 29 inclusive.*

1C. *Strategy: Find the number of pages read per week and then ...*
Each week Pat reads $1 + 2 = 3$ pages by Tuesday, $3 + 3 = 6$ pages by Wednesday, $6 + 4 = 10$ pages
by Thursday, $10 + 5 = 15$ pages by Friday, $15 + 6 = 21$ pages by Saturday, and $21 + 7 = 28$ pages
for the whole week.
METHOD 1: *Strategy: ... count forwards from the end of the 3rd week.*
In 3 weeks Pat reads $3 \times 28 = 84$ pages, with 16 pages more to read. 15 of those pages (from
page 85 to page 99) are read by Friday, as noted above. Therefore, **page 100 is read on Saturday.**

METHOD 2: *Strategy: ... count backwards from the end of the 4th week.*
In 4 weeks Pat reads $28 \times 4 = 112$ pages, finishing on a Sunday. The last page read on Saturday
was $112 - 7 = 105$ and the last page read on Friday was $105 - 6 = 99$. Thus page 100 is read on
Saturday.

*FOLLOW-UP: Investigate triangular numbers and Pascal's triangle. Locate the triangular numbers in
Pascal's triangle.*

1D. *Strategy: Handle the divisors one at a time. Start from the largest.*
The Least Common Multiple (LCM) of 6 and 5 is 30, of 30 and 4 is 60, of 60 and 3 is 60, and
of 60 and 2 is still 60. Therefore any number divisible by 2, 3, 4, 5, and 6 is divisible by 60.

METHOD 1: *Strategy: Consider the number of multiples of 60.*
Since $200 \div 60$ is 3^+ and $500 \div 60$ is 8^+, there are 3 multiples of 60 less than 200 and 8 multiples
of 60 less than 500. Thus **there are $8 - 3 = 5$ multiples of 60 between 200 and 500.**

METHOD 2: *Strategy: List the multiples of 60.*
The multiples of 60 are 60, 120, 180, <u>240, 300, 360, 420, 480</u>, 540, etc. There are 5 multiples of
60 (underlined) in the desired interval.

FOLLOW-UP: What is the first year of the twenty-first century that is divisible by 3, 6, 8, 9 and 12? [2016]

SOLUTIONS: DIVISION E

1E. _Strategy_: _Compare the areas of the two figures in two different ways._
The difference in the areas of the two noncommon regions is given as 14. If the area of the common region is added to each figure, the difference is still 14. So the area of the completed circle is 14 more than the area of the completed triangle. But the area of the circle is given as three times the area of the triangle. We are expressing the same thing two different ways.

Then three times the area of the triangle is equal to 14 more than the area of the triangle. Therefore twice the area of the triangle is 14, and **the area of the complete triangle is 7.** This last part can be visualized as follows, if each Δ represents the area of one triangle:

$$\Delta + \Delta + \Delta = \Delta + 14$$
$$\Delta + \Delta \quad = \quad 14$$
$$\Delta + \Delta \quad = \quad 7 + 7$$
$$\Delta \qquad = \quad 7$$

2A. _Strategy_: _Use the definition of multiplication._
A repeated addition of twenty 123s is equivalent to multiplying 123 by 20.
Thus, $123 + 123 + 123 + \ldots + 123 = 20 \times 123 = \mathbf{2460}$.

2B. **METHOD 1:**   _Strategy: Add Marina's distance to Lee's distance._
Since Lee travels further, he covers 2 lengths and 1 width. Marina covers 1 length and 2 widths. Together, they cover a total of 3 lengths and 3 widths; since $38 + 31 = 69$ meters, we have:

                                 3 lengths plus 3 widths = 69 meters.
Divide by three: 1 length plus 1 width = 23 meters.
Multiply by two: 2 lengths plus 2 widths = 46 meters
                             The perimeter of the pool is 46 meters.

METHOD 2:   _Strategy: Subtract Marina's distance from Lee's distance._
Both cover 1 length plus 1 width plus something. By subtraction, the second length that Lee covers is 7 more than the second width that Marina covers. Then the length and two widths that Marina covers is the same distance as (one width plus 7 meters) plus two widths. That is, three widths plus 7 meters is 31 meters. So: three widths alone is $31 - 7 = 24$ meters and the pool is 8 meters wide and $8 + 7 = 15$ meters long. The perimeter of the pool is $8 + 15 + 8 + 15 = 46$ meters.

**Follow-Up**: Four students take a test. The sum of the 3 grades other than Ana's is 260 points. The sum of the 3 grades other than Bob's is 249 points. The sum of the 3 grades other than Cara's is 255 points. The sum of the 3 grades other than Don's is 274 points. What are their test scores? [Ana - 86, Bob - 97, Cara - 91, Don - 72]

2C. _Strategy_: _Find the time each driver needs._

To drive 40 miles Sherry needs 60 minutes, so to drive 20 miles she needs 30 minutes. To drive 50 miles Terry needs 60 minutes, so for every 5 miles she drives he needs 6 minutes. Thus to drive 20 miles Terry needs $6 \times 4 = 24$ minutes. Therefore, **Terry needs $30 - 24 = 6$ fewer minutes than Sherry.**

**Follow-Up:** Aida averages 45 mph from 9 AM to 11 AM and then averages 50 mph from 11 AM to 2 PM. What is her average rate of speed for the whole trip? [48 mph, exactly]

2D. **METHOD 1:**   _Strategy_: _Separate the items bought. Eliminate one cost through comparisons._

If the same number of squares are bought using each given example, we can compare costs for the circles. Since 5 circles and 1 square cost 20¢, then 15 circles and 3 squares would cost three times as much, 60¢. But 2 circles and 3 squares cost 21¢, so those 13 additional circles cost an additional 39¢. Thus each circle costs 3¢.

Since 5 circles and a square costs 20¢, 5 circles cost 15¢ and a square costs $20 - 15 = 5$¢. Thus 12 circles cost 36¢ and 5 squares cost 25¢, and **the total cost is $36 + 25 = 61$¢.**

(Note: The same result can be obtained by purchasing the same number of circles using each given example and comparing costs for the squares.)

METHOD 2:   _Strategy_: _Manipulate the total purchases to get 12 circles._

Often it is faster to find a total directly than to find each component separately and then combine. Make the first purchase twice and the second purchase once:

First purchase:	5 circles + 1 square	costs 20¢
First purchase:	5 circles + 1 square	costs 20¢
Second purchase:	2 circles + 3 squares	costs 21¢
Total:	12 circles + 5 squares	costs 61¢

METHOD 3

Set up a table of possible costs for the circle (1, 2, 3, 4 cents, etc.) and compute the other values to see which works.

(Note: Method 1 helps set the thinking for the standard algebraic solution, which the mathlete will learn in due time. Method 3, often called brute force, is the slowest and least desirable method. However, if a student sees no other way to solve a problem, brute force is often available. We recommend that beginning problem solvers employ it without hesitation, with the expectation that they will think of more efficient methods as they gain experience.)

**Follow-Up:** At that store 2 cubes and 1 sphere costs 55¢, while 3 cubes and 4 spheres cost $1.15. How much more does a cube cost than a sphere? [8¢]

2E. *Strategy: Find one solution. Then make a change and compensate.*

Suppose all 15 darts land in the 5-ring. The total score would be the required 75 points. That is one solution.

Now suppose one dart is moved from the 5-ring to the 7-circle and another dart is moved from the 5-ring to the 3-ring. The total score would still be 75 points. This is because $5 + 5 = 3 + 7$. We now have a second way to score 75 points using all 15 darts.

All other ways to score 75 points are found by removing 2 darts at a time from the five-ring and replacing them with 1 in the three-ring and 1 in the seven-circle. Removing darts 2 at a time from the original 15 darts can be done a maximum of 7 times. Hence **there are 8 solutions.** They may be represented as (0 threes, 15 fives, 0 sevens), (1, 13 ,1), (2, 11, 2), (3, 9, 3), (4, 7, 4), (5, 5, 5), (6, 3, 6), and (7, 1, 7).

SET 3   **Olympiad 3**

3A. **METHOD 1:** *Strategy: Use the sum of their ages to consider four different ages.*

Their ages must add up to $4 \times 9 = 36$ years. Suppose Lauren is 8. Then the least sum of their ages is $8 + 9 + 10 + 11 = 38$, which is too large. Suppose Lauren is 7. Then the least sum of their ages is $7 + 8 + 9 + 10 = 34$. We need to add 2 years. Add that to either 10 $[7 + 8 + 9 + 12]$ or 9 $[7 + 8 + 11 + 10]$, or add 1 to both 9 and 10 [also $7 + 8 + 10 + 11$]. In any case, **the oldest that Lauren can be is 7 years.**

METHOD 2: *Strategy: Construct an organized table.*

Assume the three youngest sisters are 1, 2, and 3 years old; then compute the age of the oldest sister. Next, increase the ages of the three youngest sisters gradually until the age of the oldest is at its least.

$36 = (1 + 2 + 3) + 30$ $36 = (5 + 6 + 7) + 18$
$36 = (2 + 3 + 4) + 27$ $36 = (6 + 7 + 8) + 15$
$36 = (3 + 4 + 5) + 24$ $36 = (7 + 8 + 9) + 12$ [accept]
$36 = (4 + 5 + 6) + 21$ $36 = (8 + 9 + 10) + 9$ [reject]

3B. *Strategy: Use the test of divisibility for 9.*

Since the ones digit is 4, $63X904$ is even. Any multiple of 27 is also a multiple of 9 and the sum of its digits is a multiple of 9. Because the sum of the digits is $22 + X$, **X represents 5**.

Checking, $635,904 \div 27 = 23,552$, with a remainder of 0.

3C. _Strategy: Use the rules to generate the first numbers until a pattern appears._
The first ten numbers are:

1	(given)		7	(by rule 2)
2	(by rule 1)		14	(by rule 1)
4	(by rule 1)		5	(by rule 2)
8	(by rule 1)		10	(by rule 1)
16	(by rule 1)		1	(by rule 2)

Since the first and tenth numbers are the same, the machine repeats itself every 9 numbers. Thus, the 10th, 19th, 28th, 37th, and 46th numbers each are 1. Similarly, the 5th and 50th numbers are the same number. **The 50th number is 16.**

VARIATION: The 9th, 18th, 27th, 36th, and 45th numbers each are 10, so the 46th is 1, the 47th is 2, the 48th is 4, the 49th is 8, and the 50th number is 16.

FOLLOW-UP: 1998 × 1998 ends in a 4; this is called a multiplication string of two 1998s. 1998 × 1998 × 1998 ends in a 2; this is called a multiplication string of three 1998s. 1998 × 1998 × 1998 × 1998 ends in a 6; this is called a multiplication string of four 1998s. What does a multiplication string of twenty-five 1998s end in? [8]

3D. **METHOD 1:** _Strategy: Use the special distributive law._
The least common multiple of 4, 5, and 6 is 60. If 60 is added to or subtracted from any multiple of 60, the result is another multiple of 60. This is the special distributive property, and is true not only for 60, but for any number. If 6, 5, and 4 are each subtracted from 60, the results will be multiples of 6, 5, and 4 respectively. Moreover, the greater the amount that is subtracted from 60, the smaller the result is. **Therefore, the results 54, 55, and 56 are the three consecutive numbers that are divisible by 6, 5, and 4 in that order.**

METHOD 2: _Strategy: Examine the properties of all multiples of 6 and of 5._
The first number is divisible by 6, so it is even. Then the second number must be odd. It is also divisible by 5, so it ends in 5 (not 0). Thus the number divisible by 6 must end in 4. The only multiples of 6 less than 100 that end in 4 are 24, 54, and 84. 25, 55, and 85 are all divisible by 5, but of 26, 56, and 86, only 56 is divisible by 4. Therefore, 54, 55, and 56 are the three consecutive numbers.

METHOD 3: _Strategy: Compare tables of multiples, starting with those of 6._
List all the multiples of 6 less than 100. Circle those that are 1 more than a multiple of 5. From those circled, circle again those that are 2 more than a multiple of 4. Only 54 is circled twice. The three consecutive numbers are 54, 55, and 56.

METHOD 4 _Strategy: Use the fact that 4 and 6 are even._
Since the first and third numbers must be even, then the multiple of 5 is odd. List all odd multiples of 5 less than 100: 5, 15, 25, …, 95. Check the numbers before and after each to see which are multiples of 6 and 4. Only 54, 55, and 56 work.

FOLLOW-UP: What are the three least counting numbers which are divisible by all the counting numbers from 1 to 10, inclusive? [2520, 5040, 7560]

SOLUTIONS: DIVISION E

3E. **METHOD 1:**   *Strategy: Work with the cubes you don't want (Back-Door Method).*
"Unwrap" the outside painted cubes from all six faces: left, right, top bottom, front, and rear. The large cube that remains, all unpainted, is 3 by 3 by 3. **27 small cubes are not painted orange on any face.**

METHOD 2:   *Strategy: Separate the problem into parts.*
Painted on 3 faces: 8   (all the corner cubes)
Painted on 2 faces: 36   (3 cubes on each of the 12 edges)
Painted on 1 face: + 54   (9 cubes on each of the 6 faces)
                         98   (painted on one or more faces)
Then $125 - 98 = 27$ small cubes are not painted orange on any face.

FOLLOW-UPS: (1) If a cube 12 cm on each edge is painted periwinkle and then cut into 1-cm cubes, how many 1-cm cubes are painted on exactly 1 face? 2? 3? None? [600; 120; 8; 1000] (2) How can you check that your arithmetic is correct? [$12 \times 12 \times 12 = 1728$ cubes in all; the four answers above add up to 1728.]

SET 3   Olympiad 4

4A. **METHOD 1:**   *Strategy: Divide 2000 by 37 and adjust downwards.*
If 2000 is divided by 37, the remainder is 2. Thus $2000 - 2 = 1998$ is a multiple of 37. **The last year in the twentieth century is 1998.**

METHOD 2:   *Strategy: Divide 1900 by 37 and adjust upwards.*
If 1900 is divided by 37, the remainder is 13. To find the first year in the twentieth century which is divisible by 37, add $37 - 13 = 24$. Then multiples of 37 are: $1900 + 24 = 1924$, $1924 + 37 = 1961$, and $1961 + 37 = 1998$. The last year in the twentieth century is 1998.

4B. *Strategy: Work backwards.*
First, $Q + 86 + 148 = 291$, so $Q = 47$.
Then $P + 47 + 86 = 158$, so $P = 25$.
The value of P is 25.

FOLLOW-UP: Investigate Fibonacci numbers and list the first ten members of the set.

4C. **METHOD 1:** *Strategy: A group of 3 combined with a group of 2 makes a group of 5.*
Think of the class as consisting of 5 groups of equal size: 3 whole groups vote for Kim and 2 for Amy. After 8 people switch, think of the class as now consisting of 3 groups of equal size (each different in size from the original 5 groups): 2 groups vote for Amy and the other for Kim. To be divided into either 3 equal groups or 5 equal groups, the class must contain 15, 30, 45 people or some other multiple of 15.

Suppose 15 people are in the class. Then Kim would win, 9-6. Switch 8 votes and Amy would win, 14-1. Reject – the ratio is not 2:1.

Suppose 30 people are in the class. Then Kim would win, 18-12. Switch 8 votes and Amy would win, 20-10. Accept – The ratio is 2:1. Thus **a total of 30 people voted.**

METHOD 2: *Strategy: Set up two tables and see where they match:*
The number of votes Kim received is at least 8. It is also a multiple of 3.

	Kim's Votes	9	12	15	18
Actual: (use 3:2 ratio)	Amy's Votes	6	8	10	12
Switch 8 votes: (look for 2:1 ratio)	Kim's Votes – 8	1	4	7	10
	Amy's Votes + 8	14	16	18	20

Since $10 + 20 = 18 + 12 = 30$, a total of 30 people voted.

4D. *Strategy: Start with a 1-ounce stone and increase 1-ounce at a time.*
To make 1 ounce, <u>one</u> stone that weighs 1 ounce is needed.

To make 2 ounces, a <u>second</u> stone weighing 1 or 2 ounces is needed. Suppose it weighs exactly 2 ounces. Then Alma can make 1, 2, or 3 ounces by using one or both stones.

To make 4 ounces, a <u>third</u> stone that weighs up to 4 ounces is needed. Suppose it weighs exactly 4 ounces. Then Alma can make 1, 2, 3, …, 7 ounces by using one or more stones.

To make 8 ounces, a <u>fourth</u> stone that weighs up to 8 ounces is needed. Suppose it weighs exactly 8 ounces. Then Alma can make 1, 2, 3, ..., 15 ounces by using one or more stones.

To make 16 ounces, a <u>fifth</u> stone that weighs up to 16 ounces is needed. Suppose it weighs exactly 16 ounces. Then Alma can make 1, 2, 3, ..., 31 ounces by using one or more stones.
Thus, **the fewest number of stones Alma needs is 5.**

FOLLOW-UP: Investigate binary numbers (Base 2).

4E. __METHOD 1:__   _Strategy: List in an orderly way._

Group 1 (7 PATHS) From *A* through *B*:	Group 2: (7 PATHS) From *A* through *C*:	Group 3: (7 PATHS) From *A* through *D*:	Group 4: (7 PATHS) From *A* through *E*:
ABF	*ACF*	*ADF*	*AEF*
ABCF	*ACBF*	*ADBF*	*AECF*
ABCEF	*ACBDF*	*ADBCF*	*AECBF*
ABCEDF	*ACBDEF*	*ADBCEF*	*AECBDF*
ABDF	*ACEF*	*ADEF*	*AEDF*
ABDEF	*ACEDF*	*ADECF*	*AEDBF*
ABDECF	*ACEDBF*	*ADECBF*	*AEDBCF*

(Note: Since Groups 2 and 4 are mirror images of Groups 1 and 3, it is faster to count the number of paths in Groups 1 and 3, and then double that number. **There are 28 paths.)**

__METHOD 2:__   _Strategy: Use a tree diagram._

Start from *A*. Draw the four possible paths to *B*, *C*, *D*, and *E*. From each draw paths to each of the three remaining points, circling *F*. Then continue the tree diagram, circling *F* each time it appears. Use the given diagram to confirm each path. Count the number of times the circled letter *F* appears.

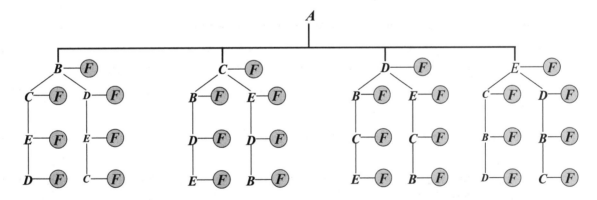

__FASTER VARIATION:__ By symmetry there are exactly as many paths that start *AC* as start *AB*, and exactly as many that start *AE* as start *AD*. Count those that start *AB* and *AD* and then double.

SOLUTIONS: DIVISION E

5A. __METHOD 1:__   *Strategy: Equate the number of figures Allen and Charlie have.*
Originally, Allen had 2 action figures more than Barbara, and Barbara 8 more than Charlie, so Allen had 10 more than Charlie. For both to have the same amount, Allen must have given 5 to Charlie. Barbara started with 8 more than Charlie, but Charlie just gained 5 action figures. Hence, **Barbara has 3 action figures more than Charlie.**

__METHOD 2:__   *Strategy: Assign convenient numbers and perform the indicated steps.*
If a problem compares some numbers without fixing their value, then assume a convenient number and watch the results. Assume Charlie starts with, say, 10. Then Barbara starts with 18 and Allen with 20. Allen gives 5 to Charlie, so that both Allen and Charlie now have 15. Then Barbara has 3 more than Charlie did. To verify the correctness of your result, repeat the above process for at least two other values for Charlie's collection. In every case, Barbara will have 3 action figures more than Charlie after the exchange.

__METHOD 3:__   *Strategy: Draw diagrams for Before and After.*
In the "Before" diagram, Charlie starts with *C* figures. This diagram shows that Allen starts with 10 more than Charlie and 2 more than Barbara. In the "After" diagram, Allen has given five figures to Charlie. It shows that Barbara has 3 more action figures than either Allen or Charlie.

5B. __METHOD 1:__   *Strategy: Compare each term to the middle term.*
The middle page number is the average of the first and ninth page numbers. It is also the average of the second and eighth, of the third and seventh, and of the fourth and sixth page numbers. The sum of all nine page numbers must be 9 times the middle page number. Since that sum is 378, **the middle page number is** $378 \div 9 = \textbf{42.}$

__METHOD 2:__   *Strategy: Compare each term to the first term.*
Add 1, 2, 3, 4, 5, 6, 7, and 8 to the number of the first page to get the numbers of the other pages. Adding all the page numbers is equivalent to increasing 9 times the first page number by $1 + 2 + 3 + 4 + 5 + 6 + 7 + 8 = 36$. Then nine times the first page number equals $378 - 36 = 342$, so that the first page number is $378 \div 9 = 38$. The middle page is $38 + 4 = 42$.

FOLLOW-UPS: *(1) Investigation: Write several sets of consecutive integers. (a) For each set, how does the middle term compare with the average? (b) Under what conditions is the average a whole number? (2) What is the average of the set of counting numbers 31, 32, 33, ..., 60 [45.5]*

SOLUTIONS: DIVISION E

5C. _Strategy: First compute D, next compute N, and finally compute the thousandth number._

Position	1	2	3	4	5	...	1000
Number		9			21	...	?

Joe needs 3 identical increases to get from 9 to 21, an interval of 12. Each increase is $12 \div 3 =$ 4. Thus the numbers in the first five positions are 5, 9, 13, 17, and 21. It takes 999 increases of 4 each to get from the first number to the 1000th number. Because $5 + (4 \times 999) = 4001$, **Joe's 1000th number is 4001.**

A neat way to simplify computation is to picture a 0th number, which would be $5 - 4 = 1$. Then there would be 1000 increases of 4 each: Since $1 + (4 \times 1000) = 1 + 4000 = 4001$, Joe's 1000th number is 4001.

VARIATION: By representing the increase by D, we can use algebra: the numbers that follow 9 are $9 + D$, $9 + 2D$, $9 + 3D$, and so on. But $9 + 3D = 21$. Then $3D$ is 12 and D is 4. Joe's 1000th number is $995 \times D$ more than the fifth number; that is $21 + 995 \times 4 = 4001$.

FOLLOW-UPS: (1) The sequence in 5C is called an arithmetic sequence. How many different ways can you find how many numbers are in the arithmetic sequence 23, 27, 31, 35, ..., 95? [19 numbers; there are many ways.] (2) In lowest terms, what are the next two terms in the arithmetic sequence $\frac{1}{6}$, $\frac{1}{4}$, $\frac{1}{3}$, ...? $[\frac{5}{12}, \frac{1}{2}]$

5D. _Strategy: Factor 111 which is_ **DDD** $\div$ **D**.

Any number of the form DDD is a multiple of 111. The sum of the digits of 111 is 3, so 111 is divisible by 3. That is, $111 = 37 \times 3$. Either $CD = 37$, or $AB = 37$, or $CD = 2 \times 37 = 74$ or $AB = 74$.

$$\begin{array}{r} A\,B \\ \times\ C\,D \\ \hline -\ -\ - \\ -\ - \\ \hline D\,D\,D \end{array}$$

(a) Suppose $CD = 37$. Then $DDD = 777$ and $AB = 777 \div 37 = 21$. Thus $CD = 37$.

(b) Suppose $AB = 37$. Then CD is a multiple of 3. Look at the units digits: $7 \times D$ ends in D. This is true only if $D = 5$. Then $DDD = 555$ and $CD = 15$.

(c) Suppose $AB = 74$. Then to produce a 3-digit result, $CD = 10, 11, 12,$ or 13 only. However, no product has the form DDD. Reject $AB = 74$.

(d) Suppose $CD = 74$. Then $DDD = 444$ and $AB = 444 \div 74 = 6$, which is not a two-digit number. Reject $CD = 74$. **Thus DDD can only be 555 or 777.**

VARIATION: Since $111 = 37 \times 3$, and AB and CD each has two digits, multiply 37 by 12, 15, 18, 21, 24, and 27. Then multiply 74 by 12 and 15. DDD can only be 555 or 777.

FOLLOW-UP: Express 1001 as the product of three prime numbers. [Use the test of divisibility for 11: $11 \times 7 \times 13$.]

SOLUTIONS: DIVISION E

5E. <u>Strategy</u>: *Examine how a total meet score can be 35 points.*

A total of $22 + 13 = 35$ points are scored. Thus either there are 5 events each distributing 7 points or 7 events each distributing 5 points. (One event worth 35 points and 35 events worth a total of 1 point each contradict given information.)

Suppose there are 7 events of 5 points each. Then the point allotment (W, L) for winning and losing is (5, 0) or (4, 1) or (3, 2). Consider each: (5, 0) is not possible because 22 points is not a multiple of 5. (4, 1) is not possible because Sandy would only score 10 points with her one win. (3, 2) is not possible because even if Candy won all 7 events, she would score less than 22 points.

Therefore there must be 5 events of 7 points each. Then the point allotment for winning and losing is (7, 0) or (6, 1) or (5, 2) or (4, 3). Consider each case: (7, 0) is not possible because 22 points is not a multiple of 7. (6, 1) is not possible because Sandy would score only 10 points with her one win. (4, 3) is not possible because even if Candy won all events, she would score less than 22 points. However, (5, 2) is possible: Candy scores $4 \times 5 + 2 = 22$ and Sandy scores $5 + 4 \times 2 = 13$. Therefore **the winner of each event gets 5 points** and the loser 2 points.

SET 4 ◆ ◆ ◆ **Olympiad 1**

1A. <u>Strategy</u>: *List 2- and 3-digit numbers separately.*

> 2-digit numbers: 15, 24, 33, 42, 51
>
> 3-digit numbers: 114, 123, 132, 141

According to conditions of the problem, 6, 60, 105, 150 must be excluded.
There are 9 numbers.

FOLLOW-UPS: Explorations (1) If we allow 0 as a digit, and also eliminate the lower and upper limits (10 and 200), what patterns occur in the solution? (2) What patterns occur if the digit-sum is 2, 3, 4, 5, etc.? [Answers will vary.]

1B. <u>Strategy</u>: *Make a table of all possible lengths and widths and compute the areas.*

The perimeter is 26, so the sum of the length and width (called the semiperimeter) is 13.

WIDTH × LENGTH	AREA
1 × 12	12
2 × 11	22
3 × 10	30
4 × 9	36
5 × 8	40
6 × 7	42

No other rectangles are possible. **The greatest area that the rectangle might have is 42 sq units.**

(Reminder: A square is a special kind of rectangle.)

FOLLOW-UP: What is the greatest whole number area if the perimeter is 18, 22, 30, 24? [20, 30, 56, 36] Can you generalize your results?

1C. _Strategy: Assume he buys and sells 12 canes in order to avoid using fractions._

12 canes cost Bryan 3 × 50¢ = $1.50.

12 canes sell for 4 × 50¢ = $2.00.

12 canes bring a profit of $2.00 – $1.50 = $.50.

Since $5.00 = $.50 × 10, multiply 12 canes by 10 = 120 canes.

Thus **Bryan must sell 120 candy canes.**

SUGGESTION: _To answer the question that was asked, always use the original words of the question in writing your answers._

1D. _Strategy: Break up the problem by layers, and make a table._

The top layer has 4 boxes. The next layer has 2 rows of 5 boxes each. The third layer has 3 rows of 6 boxes each. Make a table showing each number of rows and the number of boxes in each row.

LAYER	# of ROWS	×	# of BOXES per ROW	=	# of BOXES per LAYER
Top	1	×	4	=	4
2	2	×	5	=	10
3	3	×	6	=	18
4	4	×	7	=	28
5	5	×	8	=	40
				TOTAL =	100

In all there are 100 cereal boxes.

1E. **METHOD 1:** _Strategy: Compare the number of each type of duck to the number of sitting ducks._
Donald has 2 more normal ducks than lame ducks and 2 more lame ducks than sitting ducks, so he has 4 more normal ducks than sitting ducks. Suppose we remove 2 lame ducks and 4 sitting ducks. This gives him the same number of each type of duck. He now has a total of $33 - 2 - 4 = 27$ ducks. Thus he now has $27 \div 3 = 9$ of each type of duck.

Originally, he must have had 9 sitting ducks, $9 + 2 = 11$ lame ducks, and $11 + 2 = 13$ normal ducks. The 9 sitting ducks have 0 legs, the 11 lame ducks have 11 legs and the 13 normal ducks have 26 legs. Hence, **the 33 ducks have a total of $0 + 11 + 26 = 37$ legs.**

(Note: This method sets a firm foundation for the standard algebraic approach by utilizing the exact same thinking. Algebra is taught in the elementary grades increasingly, but not universally. This book will only offer algebraic solutions as an alternate method.)

METHOD 2: _Strategy: Form sets of "two ducks, two legs"._
Picture a group formed by pairing each sitting duck (0 legs) with a normal duck (2 legs) and set aside the extra 4 normal ducks (8 legs). In this group the numbers of legs and ducks are equal. Now picture all the lame ducks (1 leg) joining the group. The numbers of legs and ducks are still equal. With 4 ducks set aside, the group now has 29 ducks, and therefore 29 legs. Now picture the 4 extra ducks rejoining the group: it contains 33 ducks with a total of 37 legs.

METHOD 3   *Strategy: Make a table beginning with 1 sitting duck.*
The table is built using the 2-duck differences between groups. The total of 9 in the first row is 24 ducks shy of the 33 ducks, so in the second row 8 ducks are added to each group.

Sitting	Lame	Normal	Total # of Ducks
1	3	5	9
9	11	13	33

The 33 ducks have a total of $0 + 11 + 26 = 37$ legs.

SET 4   **Olympiad 2**

2A.   *Strategy: Try to duplicate each total separately.*
The minimum score is $5 \times 2 = 10$. Therefore 6 is not possible.
The maximum score is $5 \times 10 = 50$. Therefore 58 is not possible.
Since each of the target scores are even, 17 is not possible.
The remaining scores are all possible. One way is shown for each:
$14 = 2 + 2 + 2 + 2 + 4$; $38 = 10 + 10 + 10 + 6 + 2$;
$42 = 10 + 10 + 10 + 10 + 2$.
Thus, **6, 17, and 58 are not possible.**

Follow-Up: *Draw a target of two circles. Label the inner circle A and the outer ring B, with the score in A greater than the score in B. Lindsay's 5 darts scored a total of 19 points when 2 darts landed in B. How many points were assigned to ring B? [2]*

2B.   *Strategy: Do one thing at a time.*
The CD costs $20 - 6 = \$14$ and the bracelet costs $15 - 3 = \$12$, so Kelly spent $14 + 12 = \$26$.
Then \$26 spent plus \$28 remaining = \$54 to start.
Kelly had \$54 before buying the CD and the bracelet.

2C. METHOD 1:   *Strategy: Use the definition of "three times".*

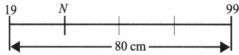

The interval from 19 to 99 is 80 cm. The diagram represents the distance from 19 to N as one part and from N to 99 as three parts. Then the entire interval from 19 to 99 consists of $1 + 3 = 4$ equal parts. Therefore each part contains $80 \div 4 = 20$ cm. Then, ***N* is $19 + 20 = 39$.** Check: from 19 to 39 is 20 cm, from 39 to 99 is 60 cm, and 60 is 3×20.

METHOD 2:   *Strategy: Use algebra.*

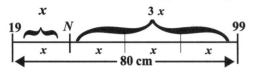

From the diagram: $\begin{aligned} x + 3x &= 80 \\ 4x &= 80 \\ x &= 20 \end{aligned}$

Then $N = 19 + x = 19 + 20 = 39$.

2D. **METHOD 1:**   *Strategy: Subtract the number of unpainted faces from the total .*

Each cube has 6 faces, so 6 cubes have 36 faces. There are 5 places where 2 cubes touch face-to-face. Thus a total of 10 faces are *not* painted blue. Then $36 - 10 =$ **26 faces of the cubes are painted blue.**

METHOD 2:   *Strategy: Separate the faces by category.*

Number of faces that are painted blue:

top faces - visible	4
bottom faces - not visible	4
vertical faces - visible	9
vertical faces - not visible	9

Thus, $(2 \times 4) + (2 \times 9) = 26$ faces of the cubes are painted blue.

METHOD 2A

Only outside faces are painted. Look at them from each of the six directions: from the top 4 faces are visible, from the bottom 4 faces, from the left 4 vertical faces, from the right 4 faces, from the front 5 faces, and from the back 5 faces. A total of 26 faces of the cubes are painted blue.

METHOD 2B

Each of the 4 cubes in the bottom layer has 4 painted faces while each of the 2 cubes in the top layer has 5 painted faces. This is a total of $16 + 10 = 26$ faces.

Follow-Up: Discussion: Suppose you use congruent cubes to make a tower. How many different configurations can you find using just 3 cubes? 4? 5? [Answers will vary depending on what limitations are placed on the way this is done. For example, is a stack of 3 different from 3 in a row horizontally?]

2E. **METHOD 1:**   *Strategy: Partition all numbers by remainder upon division by 3.*

All whole numbers leave a remainder of 0, 1, or 2 after division by 3, so we can divide them into three independent sets of numbers according to their remainders:

- Remainder 0: All multiples of 3 can be made using only 3¢ stamps: 3, 6, 9, 12, **15**, ...

- Remainder 2: <u>Use one 8¢ stamp and any number of 3¢ stamps.</u> Beginning with 8¢, 2¢ more than all multiples of 3 can be made: 8, 11, **14**, 17, etc. Amounts that cannot be made are 2 and 5.

- Remainder 1: <u>Use two 8¢ stamp and any number of 3¢ stamps.</u> Beginning with 16¢, 1¢ more than all multiples of 3 can be made: **16**, 19, 22, etc. Amounts that cannot be made are 1, 4, 7, 10 and 13.

Since amounts of 14¢, 15¢, and 16¢ can be made, every whole number that follows can be made by merely adding 3¢ stamps. **Thus, the greatest amount of postage that can't be made is 13¢.**

METHOD 2:   *Strategy: Check each amount until 3 consecutive values are found.*
Suppose we find three consecutive numbers that can be made. Then the next three consecutive numbers can also be made. All we need to do is add a 3¢ stamp to each. In this way, we can make every amount larger than the first three consecutive amounts.

Small amounts that can be made: 3¢, 6, 8, 9, 11, 12. Small numbers that cannot be made are 1¢, 2, 4, 5, 7, 10, 13. However, 14¢, 15¢, 16¢ can be made. $(14 = 2 \times 3 + 8, 15 = 5 \times 3$ and $16 = 2 \times 8$.)

Since we can make 14, 15, and 16, we can make 17, 18, and 19 by adding 3 to each. Then we can make 20, 21, and 22 by adding 3 again to each. In fact, we can make any amount above 13 by adding some multiple of 3 to 14, 15, or 16.

Thus, the greatest number which cannot be made by some combination of 3 and 8 is 13. The greatest postage which cannot be made is 13¢.

FOLLOW-UPS: (1) Suppose bolts are sold in packages of 4 and 7 each. How many consecutive numbers of bolts are needed in order to guarantee that all numbers above a certain amount can be made? What is the greatest number of bolts which cannot be bought by a combination of packages? [4,17] (2) Suppose the denominations of postal stamps are consecutive numbers such as 5¢ and 6 ¢, or 4¢ and 5¢. Find a general procedure for determining the greatest amount that can't be bought. [19¢; 11¢: Form sets of all remainders after division by the lesser denomination, and consider increasing the first multiples of the greater denomination by multiples of the lesser denomination. Other procedures also exist.]

SET 4   **Olympiad 3**

3A. *Strategy: Examine each digit separately.*
The digit "2" appears in the ones place 10 times: 2, 12, 22, ..., 92.
The digit "2" appears in the tens place 10 times: 20, 21, 22, ..., 29.
The digit "2" appears no other times. **Sandy wrote the digit "2" 10 + 10 = 20 times.**

FOLLOW-UP: A two-digit counting number is chosen at random. What is the probability that the product of its digits is odd? $[\frac{25}{90}$ or $\frac{5}{18}]$

3B. *Strategy: Work backwards from Susan's 4 pieces.*
Susan has 4 pieces.
Charles has $12 \times 4 = 48$ pieces.
Jennifer has $\frac{1}{2} \times 48 = 24$ pieces.
Paul has $\frac{1}{2} \times 24 = 12$ pieces.
Charles and Paul have 48 + 12 = 60 pieces altogether.

FOLLOW-UP: Suppose we don't know how many pieces Susan has, and suppose Charles has 90 more pieces than Paul. How many pieces would the four people have then, in all? [220]

3C. **METHOD 1:**   *Strategy: Eliminate the impossible.*

The possible products range from $1 \times 1 = 1$ to $5 \times 5 = 25$. From the 25 possibilities (1 through 25), eliminate those which are (a) prime numbers greater than 9 and (b) multiples of these primes.

       (a) Eliminate 11, 13, 17, 19 and 23:   (5 products)

       (b) Eliminate $22 = 11 \times 2$:   (1 product)

Then $25 - 5 - 1 =$ **19 products are possible.**

METHOD 2:   *Strategy: Make a table and cross out duplicates.*

Let P and Q represent the two counting numbers.

P	Q	$P \times Q$	
1	1, 2, 3, …, 9	**1, 2, 3, 4, 5, 6, 7, 8, 9**	9 products
2	1, 2, 3, …, 8	~~2, 4, 6, 8,~~ **10, 12, 14, 16**	4 new products
3	1, 2, 3, …, 7	~~3, 6, 9, 12,~~ **15, 18, 21**	3 new products
4	1, 2, 3, 4, 5, 6	~~4, 8, 12, 16,~~ **20, 24**	2 new products
5	1, 2, 3, 4, 5	~~5, 10, 15, 20,~~ **25**	1 new product

TOTAL = 19 products are possible.

FOLLOW-UP: *Two fair dice are tossed. What products are **not** possible? The solutions illustrate both categories listed in Method 1 and a third one: 32 is the product of two composites, one of which is larger than 6.*

3D. **METHOD 1:**   *Strategy: Make a table of all possible dimensions based on an area of 36.*

Find a row in which the width increases by 1 and the length decreases by 3. This occurs when we compare Row 4 against Row 3 below. Therefore **Mary's garden is 4 m wide.**

<u>Area</u>	=	<u>Width</u>	×	<u>Length</u>	
36	=	1	×	36	
36	=	2	×	18	
36	**=**	**3**	**×**	**12**	[Kevin's garden is 3 m wide.]
36	**=**	**4**	**×**	**9**	[Mary's garden is 4 m wide.]
36	=	6	×	6	[A square is a special kind of rectangle]

METHOD 2:   *Strategy: Make 2 tables; change the dimensions and then compare areas.*

Create Mary's table from each row of Kevin's table by increasing the width by 1, decreasing the length by 3, and then computing the area. Then row 3 of the table 2 is the only row in which the area is 36 sq m.

Kevin's Garden			*Mary's Garden*			
<u>W</u>	<u>L</u>	<u>Area</u>	<u>W</u>	<u>L</u>	<u>Area</u>	
1 ×	36 =	36	2 ×	33 =	66	
2 ×	18 =	36	3 ×	15 =	45	
3 ×	**12 =**	**36**	**4 ×**	**9 =**	**36**	[Mary's garden is 4 m wide.]
4 ×	9 =	36	5 ×	6 =	30	
6 ×	6 =	36	7 ×	3 =	21	

3E. **METHOD 1:** *Strategy: Track the situation, introducing one traffic light at a time.*
Suppose we call it "0 seconds when light #1 turns green. Then light #2 turns green at the 10-second mark, light #3 at the 20-second mark, and so on until light #8 turns green at the 70-second mark. The next change occurs when light #1 switches to another color at the 120-second mark. Thus **all eight light show green at the same time for 50 seconds.**

METHOD 2: *Strategy: Make a diagram.*

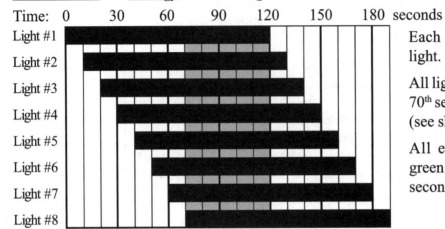

Time: 0   30   60   90   120   150   180 seconds

Light #1
Light #2
Light #3
Light #4
Light #5
Light #6
Light #7
Light #8

Each bar represents a green light.

All lights show green from the 70th second to the 120th second (see shading).

All eight lights will show green at the same time for 50 seconds.

METHOD 3: *Strategy: Make an organized chart.*
[Times On and Off = number of seconds measured from the time that Light #1 turns green]

Light #	Time On – Time Off
1	0 – 120
2	10 – 130
3	20 – 140
4	30 – 150
5	40 – 160
6	50 – 170
7	60 – 180
8	70 – 190

SET 4   Olympiad 4

4A. *Strategy: Put the greatest digits in the places with the greatest place value.*
To form the greatest product, the tens digits must be 9 and 8. That leaves 7 and 6 for the ones place. Test each arrangement: $97 \times 86 = 8342$; $96 \times 87 = 8352$. **The greatest product is 8352.**

FOLLOW-UP: Assign different sets of 4 digits. Try to discover a rule for choosing the tens digits and then a rule for matching up the ones digits with each tens digit.

4B. _Strategy_: _Select the highest available values for Gina's disks._

Gina must draw 10, 6, and 5, and Monique must draw 4, 3, and 1. **The greatest sum that Gina's disks can have is 2 + 8 + 10 + 6 + 5 = 31.** The least sum that Monique's disks can have is 7 + 9 + 4 + 3 + 1 = 24.

**FOLLOW-UPS**: (1) If Gina and Monique play a new round, can their totals be equal? [No; the sum of the ten disk numbers is 55, so one person's total is even and the other is odd.] (2) How can Gina's total be $\frac{5}{6}$ of Monique's? [Gina's sum is 25 and Monique's 30. Gina draws 2, 8, 4, 5, 6 or 2, 8, 10, 4, 1.] (3) Suppose there are 20 numbered disks, each girl selects half the disks, and Gina draws #2 while Monique draws #18. What is the greatest possible sum for Gina's disks? [139]

4C. __METHOD 1__

Strategy: _Compare the elapsed times proportionally._

At 11 AM (fast clock time), the fast clock has gained 8 minutes.

From 10 AM to 3:30 PM (fast clock time), the fast clock shows $5\frac{1}{2}$ hours of elapsed time. Therefore, at 3:30 PM (fast clock time), the fast clock has gained $5\frac{1}{2}$ hours × 8 minutes per hour = 44 minutes. Then: **The correct time is 3:30 – 0:44 = 2:46 PM.**

(Note: give credit if an answer omits the "PM")

METHOD 2:   _Strategy_: _Construct a table and note the differences in times._

Fast Clock	Correct Time	
10 AM	10:00 AM	> + 52 minutes
11 AM	10:52 AM	> + 52 minutes
12 Noon	11:44 AM	> + 52 minutes
1 PM	12:36 PM	> + 52 minutes
2 PM	1:28 PM	> + 52 minutes
3 PM	2:20 PM	> + $\frac{1}{2}$ of 52 = 26 minutes
3:30 PM	2:46 PM	

From 10:00 to 10:52 is 52 minutes. To 10:00 AM, add 52 minutes for each hour and 26 minutes for the half-hour. The correct time is 2:46 PM.

4D. **METHOD 1:**   _Strategy_: _Look for a pattern and then extend it._

Look at the last number in each row. It is a perfect square. That is, the last number in row 1 is 1 = 1 × 1; the last number in row 2 is 4 = 2 × 2; the last number in row 3 is 9 = 3 × 3; the last number in row 4 is 16 = 4 × 4; and so on. The last number in each row is the square of the row number, so the last number in row 12 is 12 × 12 = 144. Then **the first number in the 13th row is 145.**

```
          1
        2 3 4
      5 6 7 8 9
  10 11 12 13 14 15 16
   . . . And so on.
```

METHOD 2: *Strategy: Look for a pattern and then extend it.*
Look at the first number in each row: 1, 2, 5, 10, 17, …
Their differences are 1, 3, 5, 7, …
To the first number add the sum of the 12 differences:
$(1) + (1 + 3 + 5 + 7 + 9 + 11 + 13 + 15 + 17 + 19 + 21 + 23) = 1 + 144 = 145$.
The first number in the 13th row is 145.

FOLLOW-UP: Examine the pattern shown for a relationship between the	1	$= 1^2$
middle number on the left side of each row and the value on the right side	$1 + 3$	$= 2^2$
of that row. What is the sum of the first 100 odd numbers? [10,000]	$1 + 3 + 5$	$= 3^2$
	$1 + 3 + 5 + 7$	$= 4^2$, etc.

4E. *Strategy: Examine remainders for a pattern after dividing each 793 by 7.*

$$7\,)\,\overline{8\;^17\;^39\;^43\;^17\;^39\;^43\;^17\;^39\;^43\;^17\;^39\;^43\;^18}$$

The remainders after division are shown. Note that the remainders 1, 3, 4, will repeat because of the repeated division of 7, 9, and 3 by 7. After each "793" is divided by 7, the remainder is 1. Then the last division is $18 \div 7$, and the last remainder is 4. Hence, **when N is divided by 7, the remainder is 4.**

FOLLOW-UPS: (1) Divide 4,000,000,000,000,000,000 by 7. What are the remainders if the number is extended by 1, 2, 3, . . . more zeros? [5, 7, 1, 4, 2, 8, repeating] (2) What are the remainders if 4,000,000,000,000,000,000 is divided by 9 or 11? [4,;4]

SET 4   **Olympiad 5**

5A. *Strategy: Add from right to left, replacing the letters in order.*
The sum in each column is either 11 or 21.
Ones' column: $2 + 4 + D = 11$, so $D = 5$.
Tens' column: 1 (from the ones column) $+ 5 + C + 7 = 21$, so $C = 8$.
Hundreds' column: 2 (from the tens column) $+ B + 9 + 3 = 21$, so $B = 7$.
Thousands' column: 2 (from the hundreds column) $+ 6 + A = 11$, so $A = 3$.
Therefore: **$A = 3$, $B = 7$, $C = 8$, $D = 5$ (written in any order).**
(Note: If an answer does not match letter with value, then only the order 3785 is acceptable.)

```
  6 B 5 2
    9 C 4
+ A 3 7 D
---------
1 1 1 1 1
```

5B. *Strategy:* Make an organized list based upon $A + B = 10$.

Let AB be larger than BA.

AB – BA	**=**	**Difference**
91 – 19	=	72
82 – 28	=	54
73 – 37	=	36
64 – 46	=	18
55 – 55	=	0

The two numbers are 82 and 28.

FOLLOW-UPS: *(1) Select several two-digit numbers and find the difference between each and its reversal. Find the common pattern.* [Difference is always a multiple of 9] *(2) Repeat FOLLOW-UP (1) for three-digit numbers.* [Difference is always a multiple of 99]

5C. *Strategy:* Compute the missing lengths first and then compute each area.

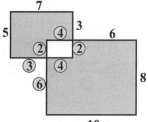

Circled lengths are computed from given lengths.

Unshaded rectangle:	length = $5 - 3 = 2$, and
	width = $10 - 6 = 4$, so
	area = $2 \times 4 = 8$
Upper rectangle:	area = $5 \times 7 = 35$
	Shaded area = $35 - 8 = 27$
Lower rectangle:	area = $10 \times 8 = 80$
	Shaded area = $80 - 8 = \underline{72}$
	Adding: 99

The sum of the areas of the shaded regions is 99 sq cm.

5D. **METHOD 1:**   *Strategy: Find the number of multiples of both 9 and 6, eliminating duplicates.*

$200 \div 9 = 22^+$: the interval $1 - 200$ contains 22 multiples of 9.

$200 \div 6 = 33^+$: the interval $1 - 200$ contains 33 multiples of 6.

The least common multiple of 9 and 6 is 18. Any multiple of 18 is also a multiple of 6 and of 9 (as well as 2 and 3).

Then $200 \div 18 = 11^+$:   the interval $1 - 200$ contains 11 multiples of 18. That is, 11 of the 22 multiples of 9 are also multiples of 6. They were counted twice. Thus, $22 + 33 - 11 = 44$. Thus, **44 numbers in the sequence are exactly divisible by either 6 or 9 or both.**

METHOD 2:   *Strategy: Make a number line.*

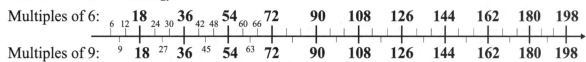

Multiples of 6:   6   12   **18**   24   30   **36**   42   48   **54**   60   66   **72**   **90**   **108**   **126**   **144**   **162**   **180**   **198**

Multiples of 9:   9   **18**   27   **36**   45   **54**   63   **72**   **90**   **108**   **126**   **144**   **162**   **180**   **198**

Every multiple of 18 appears on both lists. Cross out just one of each duplicated number and count all of the remaining numbers. Thus, 44 numbers in the sequence are exactly divisible by either 6 or 9 or by both.

(Note: Method 2 is valid but very inefficient. However, it does help to explain Method 1 more fully)

5E. _Strategy: Find the number of times 5 is a factor of the product._

For every terminal zero in our product, we need to pair a factor of 5 with a factor of 2. That is, $10 = 5 \times 2$, $100 = 5 \times 5 \times 2 \times 2$, $1000 = 5 \times 5 \times 5 \times 2 \times 2 \times 2$, and so on. In the product of the first 30 counting numbers, 5 appears as a factor 7 times: it appears once each as a factor of 5, 10, 15, 20, 25, and 30, and twice as a factor of 25. However, 2 appears as a factor of the product 26 times; it is a factor of 2, 4, 6, 8, 10, ... , 30, appearing as a factor of some numbers (such as 16) several times. Since each factor of 5 must be paired with a factor of 2, there are as many terminal zeros as there are factors of 5. To produce terminal zeroes, pair the 7 factors of 5 with 7 of the factors of 2. That is, $(5 \times 5 \times 5 \times 5 \times 5 \times 5 \times 5) \times (2 \times 2 \times 2 \times 2 \times 2 \times 2 \times 2)$. Therefore, **the product of the first 30 counting numbers has 7 terminal zeros.**

SET 5 ◆ ◆ ◆ **Olympiad 1**

1A. _Strategy: Cost it out one letter at a time, left to right._

OLYMPIADS has 9 letters. The cost is:

Letters	O	L	Y	M	P	I	A	D	S
Cost (¢)	5	5	5	6	7	8	9	10	11

The sum is 66¢.

FOLLOW-UPS: (1)How many letters does a word that costs $1.20 contain? [13] (2) Explore the relationship between this problem and triangular numbers.

1B. _Strategy: Make a chart._

Mark an **X** each time that each girl does _not_ practice.

(1) Jean does not practice at 10 AM.

(2) Krisha does not practice at noon.

(3) Lisa does not practice at 2 PM.

(4) Since Krisha practices 2 hours before Lisa, mark 📄 in the appropriate boxes to denote that **Krisha must practice at 10 AM and Lisa at noon.**
Then Jean practices at 2 PM.

Notice that when the table is completed, 📄 will appear just once in each row and just once in each column.

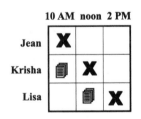

1C. **METHOD 1:**   *Strategy: Use the fact that each remainder is 5 less than the divisor.*
Notice that in each case the remainder is 5 less than the divisor. Therefore, if 5 is added to each remainder, the first list would become multiples of 6 and the second would become multiples of 11. The least common multiple (LCM) of 6 and 11 is 66. This, however, is 5 more than our desired number, which is 66 – 5 = 61. **The least counting number is 61.**

METHOD 2:   *Strategy:(Basic Approach) Make 2 lists and compare.*
Remainder = 1:   1, 7, 13, 19, 25, 31, 37, 43, 49, 55, **61**.
Remainder = 6:   6, 17, 28, 39, 50, **61**.
The least number on both lists is 61.

FOLLOW-UPS: (1) What are the next two least numbers after 61 that satisfy the conditions of the problem? [127 and 193] *(2) What are the three least counting numbers which, when divided by 4, 6, and 10, leave a remainder of 3 in each case?* [63, 123, and 183]

1D. *Strategy: First find the length of one side.*
In this Area-Perimeter problem, the connecting fact is the length of one side.
    (1) The area of each square is 96 ÷ 6 = 16 sq cm.
    (2) Then the length of one side of any square is 4 cm.
    (3) The perimeter consists of 14 such lengths.
    (4) **The perimeter is** 14 × 4 = **56 cm**.
(Note: Because of the common sides, all of the squares must be congruent to each other.)

1E. The row with the broken chair is counted twice (third row from the front and eighteenth row from the back), so the auditorium has 3 + 18 – 1 = 20 rows. In this row, the broken seat is not counted, so the row has 12 + 17 + 1 = 30 seats. Thus 20 rows of 30 seats each, or **600 seats, are in the auditorium.**

FOLLOW-UP: If each of the 20 rows has one seat more than the row in front of it and the first row has 20 seats, how many seats are in the auditorium? [590]

SET 5   **Olympiad 2**

2A. **METHOD 1:**   *Strategy: Combine operations.*
If you multiply by 4 and then divide by 12, that is equivalent to dividing by 3. If her cat's age is divided by 3, the result is 5, so **her cat's age is 15.**

METHOD 2:   *Strategy: Work backwards.*
A number divided by 12 is 5, so the number is 60. Then her cat's age multiplied by 4 is 60, so her cat's age is 15.

2B. _Strategy: Remove each coin, one at a time and check for divisibility by 7._
The total value of all the coins is $1.90. One coin can be removed in 5 different ways:

Coin removed	nickel	dime	quarter	half-dollar	dollar
Sum of remaining coins	1.85	1.80	1.65	**1.40**	0.90

Only $1.40 is divisible by 7. **Amy lost the half-dollar.**

2C. _Strategy: Proceed one sentence at a time._
(1) The number is 1 2 _ _ or 2 1 _ _ or 3 0 _ _.
(2) The number is 1 2 8 _ or 2 1 4 _ or 3 0 0 _. Reject 3 0 0 _ (2 equal digits).
(3) The number is 1 2 8 8 or 2 1 4 9. Reject 1288 (2 equal digits).
The number is 2149.

2D. **METHOD 1**

The length of the rectangle is also the length of one side of the square ("side"). The width of the rectangle ("$\frac{1}{2}$ side") is half the length of one side of the square.

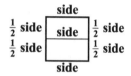

Therefore the perimeter of the rectangle (given as 24 cm) is $\frac{1}{2} + 1 + \frac{1}{2} + 1 = 3$ times the length of one side of the square ("3 × side"). Then the length of one side of the square is 24 ÷ 3 = 8 cm, and **the area of the square is 64 sq cm.**

METHOD 2: _Strategy: Draw a convenient line._
Fold the square paper again, perpendicular to the first fold. This yields 4 congruent smaller squares. Call the side-length of each small square one unit. Then the perimeter, given as 24 cm, is also equal to 6 units. Thus one unit is equal to 4 cm, one side-length for the square is 8 cm and the area of the square is 64 sq cm.

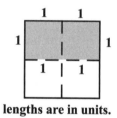

lengths are in units.

METHOD 3: _Strategy: Use algebra._
To avoid the use of fractions, let $2s$ represent the length of one side of the square as shown. Then the perimeter of one of the congruent rectangles is $6s$ or 24. Thus $s = 4$, and the length of one side of the square is $2s$ or 8. The area of the square is 64.

FOLLOW-UPS: _Extend the question: what is the area if the paper is folded in thirds? quarters? fifths? To gain integral values for the length of one side, what relationship must exist between the number of rectangles and the given perimeter? To foster greater depth of understanding, what perimeter must be assigned if the paper is folded into 100 rectangles? [81 sq cm; 92.16 sq cm; 100 sq cm; the semiperimeter must be a multiple of one more than the number of rectangles; any multiple of 202 cm]_

2E. From 1:00 to 1:30, the car from Boston travels 25 miles. There are 270 miles left for the two cars to travel. From 1:30 until they pass each other, the car from Boston travels 50 miles each hour while the car from New York travels 40 miles each hour. Together they travel a total of 90 miles toward each other each hour. Thus, to travel 270 miles they need 3 hours starting at 1:30. **The cars pass each other at 4:30.**

FOLLOW-UP: Suppose both cars traveled south to Miami instead and followed the same rates, starting times, and routes as before, and neither made any stops. (1) At what time would they meet? (2) How far would each travel? (3) Consult a map. Would they meet before they reach Miami? Where? [4:30 PM – but a day later! They would meet 27 hours after the second car leaves. The cars would meet 1080 miles south of New York, near the Kennedy Space Center, about 200 miles north of Miami.]

SET 5   **Olympiad 3**

3A. **METHOD 1:**   *Strategy: Make a simpler problem.*
Since 9 widgets cost $90, one set of 3 widgets cost $30. The third of these 3 widgets cost $4, so two regularly priced widgets cost $26. **One costs $13.**

METHOD 2:   *Strategy: Represent the problem visually.*
Let each ⬜ represent the regular price of a widget.

1. ▭ + $4   ▭ + $4   ▭ + $4   costs $90.
2. ▭   ▭   ▭   costs $78.
3. ⬜   costs $13.

(Note: Many children understand a problem much more fully if it is presented visually.)

METHOD 3:   *Strategy: Remove the "sale" widgets.*
Since I buy 9 widgets, I pay for 6 at the regular price and for 3 at $4 each. For the 6 widgets at regular price, I pay a total of $90 − 12 = $78. Thus the regular price of a widget is $78 ÷ 6 = $13 each.

3B. **METHOD 1:**   *Strategy: Examine when the next reversal might occur.*
Newton's father is 27 years older than Newton. His father is 41 now. The next possible reversal of the digits could occur when his father is in his fifties and Newton is in his twenties. That is, at 52 and 25 years of age. Since the difference is still 27 years, the problem is solved. **Their ages again have their digits reversed eleven years later.**

SOLUTIONS: DIVISION E

METHOD 2   *Strategy: Make a table that keeps the age difference constant.*
List various ages for Newton and add 27 years for his father's age. Look for reversed digits.

Years later	1	2	3	4	5	6	7	8	9	10	11
Newton's age	14	15	16	17	18	19	20	21	22	23	24	25
Father's age	41	42	43	44	45	46	47	48	49	50	51	52

Eleven years later the digits in their ages are reversed again.

Follow-Ups: *(1) Suppose each lives to be 80 years old. How many times will their ages have their digits reversed while both are alive?* [4 times] *(2) Examine the differences between several pairs of reversed ages.* [The differences will always be a multiple of 9.]

3C. METHOD 1:   *Strategy: Use the number of minutes the clock loses.*
From 1 PM to 10 AM is 21 hours. Since the clock loses 3 minutes every hour, it loses a total of 63 minutes = 1 hour 3 minutes during the 21 hours. **At 10 AM the clock shows 8:57.**

METHOD 2:   *Strategy: Use the number of minutes the clock actually shows.*
The clock shows 57 minutes for each true hour that passes. In 21 hours, the clock advances $21 \times 57 = 1197$ minutes. Since 1200 minutes = 20 hours, the clock advances 3 minutes less than 20 hours, to 8:57.

Follow-Up: *How many hours will elapse before the clock again shows the right time?* [240 hours]

3D. *Strategy: Use the equality of the areas to find the length of the rectangle.*
The area of the square is 36 sq cm. Since the rectangle has the same area as the square, the area of the rectangle is also 36 sq cm. The length of the rectangle is $36 \div 3 = 12$ cm. Hence, **the perimeter of the rectangle is** $3 + 12 + 3 + 12 =$ **30 cm.**

3E. *Strategy: Find the number of numbers. Then minimize all but the required number.*
Suppose a different set of 3 numbers has a sum of 15. Then its average is $15 \div 3 = 5$. It is also true that $3 \times 5 = 15$ and that $15 \div 5 = 3$. Therefore, if the sum of a set is divided by its average, the result is the number of numbers in the set. In this case, the set contains $350 \div 50 = 7$ numbers.

Because one of those numbers is 100, the sum of the other six numbers is 250. To maximize the required number, minimize the remaining five numbers. Those five counting numbers are 1, 2, 3, 4, and 5, with a sum of 15. Then **the greatest number is** $250 - 15 =$ **235.**

Follow-Up: *The sum of several counting numbers is 100 and their average is 20. If three of the numbers are 25, 30, and 35, and no two numbers are equal, list all sets of numbers that are possible.* [The rest of the 5-member set are 1 and 9, 2 and 8, 3 and 7, or 4 and 6.]

4A. _Strategy:_ _Work from right to left._

A		7						e	d	c	7	b	4

Since the sum of 7, b and 4 is 20, $b = 9$. Then, the sum of 9, 7 and c is 20, so that $c = 4$. Continuing this process from right to left one box at a time, $d = 9$, $e = 7$, and so on. Then **the value of A is 9**. The number turns out to be 94,794,794,794,794. The leftmost 7, which was given, serves as a check.

FOLLOW-UP: In problem 4A, to maintain a constant sum of 20, the two digits of 7 and 4 are given. To make a new problem with the constant sum of 20, what digits could not replace either the 7 or the 4? [0, 1, 2: If two of the digits are 9 and 8, then the smallest usable digit is 3.]

4B. _Strategy:_ _Compute the income and expenses separately._
The grocer spends \$15 for 180 oranges. After throwing away the 20 rotten oranges, she sold the remaining 160 oranges. This is 20 sets, each with 8 oranges. She grossed $20 \times 85\cent = \$17.00$. **Her profit is** $\$17 - \$15 = \$2.$

4C. _Strategy:_ _Consider all possible dimensions of the base._
Because the perimeter of the base is 18 cm, the _semiperimeter_ (length + width) is 9 cm. Each edge of every cube measures 1 cm, so the dimensions of the base of the rectangular solid are 1×8, 2×7, 3×6, or 4×5.

There are 42 cubes so the product of the length, width, and height of the rectangular solid is 42. The dimensions of the base are both factors of 42. Only 2×7 of the possible dimensions above produces a whole number factor of 42. Then, with a base whose dimensions are 2 cm by 7 cm, **the height of the brick is 3 cm.**

4D. **METHOD 1:**   _Strategy:_ _Use the fact that division is the inverse operation of multiplication._
The value of A cannot be 0 because A is a leading digit. The value of B is either 5 or 0. It cannot be 0 because if it were, then the final product would end in 40; however, 2730 ends in 30. Therefore, **the value of B is 5.**

$$\begin{array}{r} A\ 8 \\ \times\ \ 3\ B \\ \hline 2\ 7\ 3\ 0 \end{array}$$

Since $A8 \times 35 = 2730$, $2730 \div 35 = A8$.

The quotient is 78, so **the value of A is 7.**

Check by multiplying 78 by 35.

METHOD 2   *Strategy: Use the process of elimination to find the value of A.*
The ones digit in the product is 0, so the value of *B* is 5, as above.
The first two digits in the product are 27, so *A* cannot be 1, 2, 3, 4,
5, or 6. Nor can *A* be 9, because 3 × 9 plus the carry would be more
than 27. Then replace *A* by 7 and 8 to see which works.

$$
\begin{array}{r} 7\ 8 \\ \times\ \ 3\ 5 \\ \hline 3\ 9\ 0 \\ 2\ 3\ 4 \\ \hline 2\ 7\ 3\ 0 \end{array}
\qquad
\begin{array}{r} 8\ 8 \\ \times\ \ 3\ 5 \\ \hline 4\ 4\ 0 \\ 2\ 6\ 4 \\ \hline 3\ 0\ 8\ 0 \end{array}
$$

METHOD 3   *Strategy: Use the Commutative Principle.*
The ones digit in the product is 0, so the value of *B* is 5, as above.
Rewrite the example, switching the multipliers as shown at right.
Since 28 + *EDC* = 273, *EDC* = 245. Then *A* = 245 ÷ 35 = 7.

$$
\begin{array}{r} 3\ 5 \\ \times\ \ A\ 8 \\ \hline 2\ 8\ 0 \\ E\ D\ C \\ \hline 2\ 7\ 3\ 0 \end{array}
$$

4E.   METHOD 1A: *Strategy: Combine all the climbing. Represent visually.*
From the diagram at the right the top of the ladder is 11 rungs above the
middle rung. Then the bottom of the ladder is 11 rungs *below* the middle
rung. **The entire ladder has a total of** 11 + 1 + 11 = **23 rungs.**

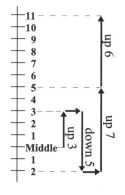

METHOD 1B: *Strategy: Combine all the climbing.*
Replace the diagram in Method 1A with the following: Starting at the
middle and ending at the top, the following movements were made:
3↑ + 5↓ + 7↑ + 6↑ = 16↑ + 5↓ = 11↑,
where ↑ denotes movement up the ladder and ↓ denotes movement down
the ladder. The rest of Method 1A is unchanged.

METHOD 2:   *Strategy: Assume the simplest numbers and then adjust.*
Suppose there are, say, 5 rungs. Then the middle rung is the 3rd. From there the firefighter
proceeds, in order, to the 6th rung, the 1st rung, the 8th rung, and finally the 14th rung. From the 3rd
rung (assumed as the middle) to the 14th rung (assumed as the top) is 11 rungs. Then there are 11
rungs above the middle and 11 below the middle for a total of 23 rungs, and the middle is really
the 12th rung.

FOLLOW-UPS: (1) Suppose the middle rung was the 100th. How many rungs would that ladder have?
[199: 99 above, 99 below]. *(2) If we have a middle rung, is it possible to have an even number of rungs?*
[No] *(3) [Research project] How is this problem related to problems involving the median of a set of
numbers?*

SET 5   **Olympiad 5**

5A.   METHOD 1:   *Strategy: Compare Tony's purchases to Joanie's.*
Tony has 2 more candy bars than Joanie, but he paid 50¢ more. Each candy bar costs 25¢.
Joanie started with 25 + 70 = 95¢.

METHOD 2   *Strategy: Draw a picture.*

Tony's candy and money   Joanie's candy and money

| | | | + 20 | = | | + 70 |

| | | = 50 | (Remove 1 bar and 20¢ from each person) |

| | = 25 | (Divide by 2) |

Joanie started with 25 + 70 = 95¢.

METHOD 3   *Strategy: Use algebra.*

Let C = the price of 1 candy bar, in cents.
Then 3C = the price of 3 candy bars.
Joanie starts with C + 70 cents.
Tony starts with 3C + 20 cents.

$$3C + 20 = C + 70$$
$$3C = C + 50$$
$$2C = 50$$
$$C = 25$$ Joanie started with 25 + 70 = 95¢.

FOLLOW-UP: Suppose stamps now cost 33¢ for the first ounce and 22¢ for each additional ounce. Before, they cost 32¢ for the first ounce and 23¢ for each additional ounce. For how many ounces do stamps now cost 6¢ less than they did before? At these rates, for how many ounces would stamps cost $1.00 less now than before? [8, 102]

5B. The man binds 100 books and his helper binds 25 books for a total of 125 books every 2 days. Thus, binding 500 books requires 4 sets of 2 days each. **It would take 8 days to bind 500 books.**

5C. *Strategy: Use the power of place value.*

In the hundreds column, place the largest 3 choices: 9, 8, 3 in some order. It follows that, in the tens column, $A = 1$. Note that in the ones column, D appears twice, M appears once and B does not appear. Then $D = 9$, $M = 8$ and $B = 3$. Substituting the values, $319 + 819 + 918$. **The greatest possible sum is 2056.**

$$\begin{array}{r} BAD \\ MAD \\ + DAM \end{array}$$

FOLLOW-UPS: (1) What is the least possible sum? [1475] (2) If the choices are 1,3,5,and 7 and BAD + MAD + DAM = 1267, what is the value of the four-digit number ABDM? [5173]

5D. *Strategy: Draw a diagram.*

As shown, the front of house *ABCD* would be $70 - 10 - 10 = 50$ feet.
The side of house *ABCD* would be $100 - 20 - 30 = 50$ feet.
The largest area the house can have is $50 \times 50 =$ 2500 sq. ft.

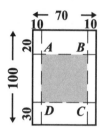

5E. *Strategy: Represent the problem visually.*

class | passing | class | passing | class | passing | class

The elapsed time from 8:26 to 11:26 is 3 hours, which is 180 minutes. There are 3 (not 4) passing periods between classes, for a total passing time of 12 minutes. This leaves a total of 168 minutes for the four classes. **One class period contains 42 minutes.**

FOLLOW-UP: (A classic problem) Farmer Grey builds a fence for the front of the South Field, a total of 90 meters. Each fence section is 3 m long and costs $20. The sections are placed end-to-end. Each end of every section is then nailed to a fence post. If each fence post costs $3, how many dollars does it cost to build the fence? [$693; 30 sections and 31 posts are needed.]

SET 6 ◆ ◆ ◆ Olympiad 1

1A. **METHOD 1:** *Strategy: Compare similar numbers.*

Notice that each number being added is 100 more than one of the numbers being subtracted.
The value is $100 + 100 + 100 =$ 300.

METHOD 2: *Strategy: Group by operation.*

Add the numbers $268 + 1375 + 6179 = 7822$. Then add the numbers $168 + 1275 + 6079 = 7522$.
Finally, subtract the totals: $7822 - 7522 = 300$.

1B. *Strategy: Draw a Picture.*

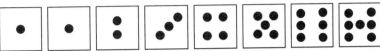

Draw 8 boxes. Then put the least possible number of marbles in each box. Put 1 marble in box 1. Then put 1 marble in box 2. You can't have 3 boxes with the same number of marbles, so put 2 marbles in box 3, 3 marbles in box 4, and so on. **The fewest marbles is** $1 + 1 + 2 + 3 + 4 + 5 + 6 + 7 = $ **29 marbles.**

1C. *Strategy: Consider one statement at a time. Eliminate numbers which do not satisfy all conditions.* Numbers which are multiples of both 3 and 5 are the multiples of 15. Begin with the set of whole numbers and then pare it down by considering the given conditions consecutively.

Condition	Whole numbers that satisfy all conditions
Less than 100	0, 1, 2, 3, …, 99
Multiples of 15	0, 15, 30, 45, 60, 75, 90
Odd	15, 45, 75
Sum of digits is odd	45

The number is 45.

1D. METHOD 1: *Strategy: Include a 0¢ stamp and count the number of combinations.*
  Number of choices for 1¢ stamps, including 0, is 5: (0,1,2,3,4).
  Number of choices for 5¢ stamps, including 0, is 4: (0,1,2,3).
  Number of choices for 25¢ stamps, including 0, is 4: (0,1,2,3).
Combine each 1¢ choice with each of the 5¢ choices, and then combine each result with each of the 25¢ choices. Therefore, the total number of choices for all stamps is $5 \times 4 \times 4 = 80$. However, 80 includes the choice of having no stamps at all. Since it is given that every choice must be 1¢ or greater, there are 79 amounts of at least 1¢.

METHOD 2: *Strategy: Separate the stamps into three groups.*
(a) Four amounts can be formed by using 1¢ stamps only: 1, 2, 3, 4.
(b) Fifteen amounts can be formed by using 5¢ and 25¢ stamps only:
   5, 10, 15; 25, 30, 35, 40; 50, 55, 60, 65; 75 ,80 , 85, 90.
(c) There are too few 1¢ and 5¢ stamps to make any amounts in two different ways. Each of the four amounts in (a) can be combined with each of the fifteen amounts in (b), thus producing $15 \times 4 = 60$ different amounts that combine all three types of stamps.

   Total Amounts: (a) 4
                  (b) 15
                  (c) 60
   GRAND TOTAL: **79 different amounts.**

METHOD 3 *Strategy*: *Establish a maximum and then eliminate all impossibilities.*
Find the largest postage amount that can be made with the stamps, and then subtract the number of smaller values that cannot be made.

The maximum that can be made is $4 + 15 + 75 = 94¢$. The 15 values less than $94¢$ that cannot be made are those that require four $5¢$ stamps. These are 20, 21, 22, 23, 24, 45, 46, 47, 48, 49, 70, 71, 72, 73, and $74¢$. The number of possible amounts is $94 - 15 = 79$.

FOLLOW-UPS: *(1) Alana has 3 pennies, 2 nickels, 1 dime, and 2 quarters. How many different amounts of at least 1 cent can she make with these coins? [59] (2) How many whole numbers are divisors of 72? [12]*

1E. *Strategy*: *Find the length of one side of the figure.*
Because of the common sides, all the squares are congruent to each other. The perimeter consists of 12 equal sides. The length of a side is $72 ÷ 12 = 6$ cm. The area of each square is $6 × 6 = 36$ cm^2. **The area of the figure is $5 × 36 =$ 180 cm^2.**

FOLLOW-UPS: *(1) The area of the given figure is 150 cm^2. Find its perimeter. [60 cm] (2) The area and perimeter of the figure shown are numerically equal to each other. What is the perimeter? [24 units]*

SET 6 | **Olympiad 2**

2A. *Strategy*: *Utilize one concept at a time to eliminate possibilites.*
Any number between 301 and 370 has a hundreds digit of 3. The sum of the digits is $5 × 3$ or 15. Then the sum of the tens digit and the units digit is $15 - 3$ or 12. The number is odd, so the units digit is odd. Test 339, 357, 375, and 393. 339 does not have 3 different digits. 375 and 393 are greater than 370. Only 357 satisfies all the conditions. **The number is 357.**

FOLLOW-UPS: *(1) An __even__ number between 301 and 370 has three different digits. If the sum of the digits is five times the hundreds digit, find the number. [348] (2) Suppose the sum of the digits is 4 times the hundreds digit. Can the students change another condition of the problem to produce a unique answer? [Yes; make two digits the same.]*

2B. **METHOD 1:** *Strategy*: *Make a much simpler problem.*
If we subtract 2 from $47Y6$, the difference will be 2 greater than 1998. Then $47Y4 + 2000 = PR5T = 67Y4$. $P = 6$, $R = 7$, $T = 4$, $Y = 5$. Therefore, **$PR5T =$ 6754.**

$$\begin{array}{r} P\,R\,5\,T \\ -\ 4\,7\,Y\,4 \\ \hline 2\,0\,0\,0 \end{array}$$

METHOD 2:   *Strategy: Change subtraction into addition.*

The inverse of subtraction is addition. Add the difference and the subtrahend to find the minuend. Start with the ones: $8 + 6 = 14$, so $T = 4$. Then $9 + 1 + Y$ ends in 5, so $Y = 5$. Since $R = 7$ and $P = 6$, the number $PR5T$ represents 6754. Check: $6754 - 4756 = 1998$.

$$\begin{array}{r} 1\,9\,9\,8 \\ +\ 4\,7\,Y\,6 \\ \hline P\,R\,5\,T \end{array}$$

2C.   METHOD 1:   *Strategy: Place the numbers around a circle.*

Since the first person in line goes to the end of the line, the line of people can be thought of as a circle of people. Write the numbers 1 through 10 around a circle. Beginning with the number 2, eliminate every second number. In order, eliminate 2, 4, 6, 8, 10, 3, 7, 1, 9. The last remaining number is 5. **The last remaining person occupied the 5th position originally.**

METHOD 2:   *Strategy: Use 10 numbered slips to act out the movements.*

$1\,2\,3\,4\,5\,6\,7\,8\,9\,T$	[10 slips are in line. **1** is in front, and **T** (for 10) is in back.]
$3\,4\,5\,6\,7\,8\,9\,T\,1$	[**1** goes to the back and **2** sits down.]
$5\,6\,7\,8\,9\,T\,1\,3$	[**3** goes to the back and **4** sits down.]
$7\,8\,9\,T\,1\,3\,5$	[And so on ...]
$9\,T\,1\,3\,5\,7$	
$1\,3\,5\,7\,9$	
$5\,7\,9\,1$	
$9\,1\,5$	
$5\,9$	
5	The last remaining person was 5th in the original line.

FOLLOW-UP: *Suppose the line originally contains 2, 3, 4, 5, . . . people. What is the pattern to the position of the last remaining person?* [1, 3, 1, 3, 5, 7, 1, 3, 5, 7, 9, ...] *How many people are on line originally if the person in position 1 is the last remaining person?* [1, 2, 4, 8, 16, ...] *Why is it related to powers of 2?* [Each pass through the line eliminates half the people. Person 1 is the last remaining if the number of people after each pass is still even.]

2D.   METHOD 1:   *Strategy: Find the total surface area of the cube.*

A 10-cm cube has a surface area of 100 square centimeters on each of its 6 faces, making a total of 600 sq cm to be covered. To get an area of 600 sq cm of tape which is 2 cm wide, divide 600 sq cm by 2 cm. **The length of tape needed to cover the entire cube is 300 cm of tape.**

METHOD 2:   *Strategy: Find the length of tape needed to cover one face of the cube.*

One face of a cube is 10 cm by 10 cm. To cover it with 2-cm tape, 5 pieces are needed, each 10 cm long. That is a total of 50 cm of tape. Since a cube has 6 faces, the length of tape needed to cover the entire cube is $6 \times 50 = 300$ cm of tape.

FOLLOW-UP: *What is the length of the tape needed if its width is 1 cm, $\frac{1}{2}$ cm, $\frac{1}{4}$ cm, and so on?* [600, 1200, 2400, etc.] *Why does the length double when the width is halved? This FOLLOW-UP together with Method 1 can be used to explain division by a fraction.*

2E. **METHOD 1:** *Strategy: First determine how much each new member paid.*
When 3 new members joined the club, the five original members each saved $15, for a total of $75. Thus the three new members paid a total of $75, which was $25 each. Since, therefore, each of the eight members paid $25, **the price of the used computer was $200.** As a check, the five original members would have paid $40 each for a total of $200.

METHOD 2: *Strategy: Find the multiples of 5 and 8.*
Since both five people and eight people can share the cost equally, the total cost has to be a multiple of both 5 and 8. For the least common multiple, 40, the savings is $15 \div 5 = \$3$ per person. To get savings of 5 times that amount, multiply 40×5 for an original price of $200. Check: Five people sharing the cost equally is $200 \div 5 = \$40$. Eight people sharing the cost equally is $200 \div 8 = \$25$. Each original member saves $40 - 25 = \$15$.

METHOD 3: *Strategy: Find the multiples of 5 and 8.*
The computer costs 5 "old shares" or 8 "new shares".
Each new share is $15 less than an old share.
In dollars, 5 old shares = 8 old shares − 120 (the result of $8 \times \$15$).
3 old shares = $120.
1 old share = $40.
The computer costs 5 old shares or $200.

FOLLOW-UP: Five members of a computer club buy a computer, dividing the cost equally. They each contribute $40. Later, one member leaves and wants her money back so the remaining members return her share. How much additional money does each member have to contribute? [$10]

3A. *Strategy: Proceed one condition at a time.*
The thousands digit can't be 2 or greater because then the ones digit would be more than 9. Thus, the thousands digit is 1. Also, the question reads, "Dr. Bolton *WAS* born...".

Condition	Number
The tens digit is twice the thousands digit	1 _ 2 _
The ones digit is 3 times the tens digit	1 _ 2 6
The hundreds digit is the sum of the other three digits	1 9 2 6

Dr. Bolton was born in 1926.

SOLUTIONS: DIVISION E

3B. **METHOD 1:** *Strategy: Count in an organized way.*
To make a cube measuring 5 cm on each side, you need
$5 \times 5 \times 5 = 125$ one-cm cubes. Count the cubes in the staircase
by slicing the stairs vertically (*Figure 1*) into 5 congruent slices.
Each slice contains 15 cubes. There are $15 \times 5 = 75$ cubes in
the staircase. **You need $125 - 75 = 50$ more one-cm cubes.**
Alternatively, count cubes using horizontal layers (*Figure 2*).

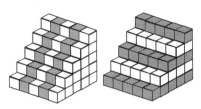

Figure 1   *Figure 2*

METHOD 2: *Strategy: Complete the cube.*
The staircase already measures 5 cm in height, width, and depth. Add cubes to the staircase
layer by layer to change it into a cube, as shown in *Figure 3* and the table below.

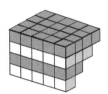

Figure 3

Layer	Original number of cubes	Additional number of cubes
Bottom	25	0 more
Second	20	5 more
Third	15	10 more
Fourth	10	15 more
Top	5	20 more

A total of $5 + 10 + 15 + 20 = 50$ more cubes were added to the
staircase to make it into a cube.

3C. *Strategy: Make an organized list.*
List the numbers that have their digits in decreasing order. Group them by hundreds.

Grouping	Numbers whose digits decrease	Quantity
100s	——	0
200s	210	1
300s	321, 320, 310	3
400s	432, 431, 430, 421, 420, 410	6
500s	543, 542, 541, 540, 532, 531, 530, 521, 520, 510	10
	Total	20

There are 20 whole numbers between 100 and 599 whose digits are in decreasing order.

FOLLOW-UPS: *(1) How many such numbers are there between 600 and 699? [15] What pattern do you
notice? [1, 3, 6, 10, ... = 1, 1 + 2, 1 + 2 + 3, 1 + 2 + 3 + 4, ...] (2) Explore triangular numbers.*

3D. **METHOD 1** *Strategy: Consider how frequently both lights flash together.*
From 1:00 to 3:00 is 120 minutes. We need the least multiple of 14 that is greater than 120. 120
divided by 14 is between 8 and 9. Then $9 \times 14 = 126$ is the least such multiple of 14. 126 minutes
after 1:00 PM is 3:06 PM. **The first time they flash together after 3 PM is 3:06 PM.**

SOLUTIONS: DIVISION E

METHOD 2:   *Strategy: List all times.*
Both lights flash together at 1:00 PM. They then flash together at: 1:14, 1:28, 1:42, 1:56, 2:10, 2:24, 2:38, 2:52, and the time we want, 3:06 PM.

3E.   **METHOD 1:**   *Strategy: Start with the relationship between the dimes and pennies.*
The number of pennies is twice the number of dimes. Then the combined value of the pennies and dimes is a multiple of 12¢: 12, 24, 36, 48, 60, 72 cents.

The difference between 76¢ and one of the above values must be a multiple of 5. Only 36¢ satisfies this condition: 76¢ – 36¢ = 40¢. **Richard has 8 nickels.** (Surprise! We never needed to know that he had 17 coins!)

METHOD 2:   *Strategy: Start with the relationship between pennies and 76¢.*
To make 76¢, Richard must have 1 or 6 or 11 or 16 or … pennies. Since the number of pennies is twice the number of dimes, the number of pennies is even; he has 6 or 16 or 26 or … pennies. He cannot have 16 or more pennies, or else he would have at least 8 dimes with a total value greater than 76¢. Thus, he has 6 pennies and 3 dimes. The remaining 8 coins are nickels, so Richard has 8 nickels. Check : 6¢ in pennies, 40¢ in nickels, and 30¢ in dimes add up to 76¢.

METHOD 3:   *Strategy: Consider the number of dimes among the 17 coins.*
Make a table based on the number of dimes (because dimes have the greatest value) and the fact that the total number of coins is 17. Look for a total value of 76¢.

Dimes	Pennies	Nickels	Value
1	2	14	82¢
2	4	11	79¢
3	6	8	76¢

Richard has 8 nickels.

FOLLOW-UP: Tara has coins with a value of $1.37. The coins are pennies, nickels, and dimes. She has 4 times as many pennies as dimes. What is the least number of nickels Tara could have? [5]

4A.   Every 7 hours the hand returns to its starting point. After 7, 14, and 21 hours, the hand points to 5. After 24 hours, the hand has moved 3 numbers past 5, and **24 hours from now, the hand points to 1.**

FOLLOW-UPS: (1) If today is Friday, what day of the week will it be 24 days from now? 240 days? [Monday; Sunday] *(2) EXPLORATION: Experiment by replacing 5 and/or 24 with other numbers. What conclusions can students draw?*

4B. *Strategy: Add or subtract lengths of sides of the square to find the lengths of segments.*
The sides of the small squares are each 3 m and 4 m, and of the large square, 7 m. **The perimeter of the shaded region is** $7 + 4 + 3 + 1 + 4 + 3$
= 22 m.

The shaded region is bordered by 6 segments ("walls"). Notice that the combined length of the two bottom walls (4 m + 3 m) equals that of the top wall (7 m) and that the combined length of the two left walls (1 m + 3 m) equals that of the right wall (4 m). Moving the 4 m wall on the bottom of the shaded region down by one meter changes the shape of the region into a rectangle, whose perimeter does not change. Therefore, this problem could have been solved by finding the perimeter of a 7m by 4 m rectangle.

4C. **METHOD 1:** *Strategy: Increase each number by 1 to increase the average by 1.*
Suppose the seventh number is 4. Then the average of the seven numbers is still 4. To change the average to 5, add 1 to each of the seven numbers. Increasing the total by 7 is equivalent to adding 7 to the new number. Thus, **the seventh number is** $4 + 7 = $ **11.**

METHOD 2: *Strategy: Use the definition of an average.*
The sum of the original 6 numbers is 6×4 or 24. When the new number is added, the sum is 7×5 or 35. The seventh number is $35 - 24 = 11$.

FOLLOW-UP: Suppose 2 more numbers are added to the original 6. If the average of the 8 numbers is 10, what is the average of the two numbers that were added? [28]

4D. *Strategy: Fill in the squares one at a time.*
Step 1: From row 3, we see that C can't be 4 or 1. From column 2, C can't be 2. Therefore C must be 3.
Step 2: From row 3, D must be 2.
Step 3: From column 4, we see that E can't be 4 or 1. From the diagonal $1FDE$, E can't be 1 or 2, since D is 2. Then E must be 3.
Step 4: From the diagonal, F must be 4.
Step 5: From column 2, B must be 1.
Step 6: From column 1, A can't be 1 or 4. From row 4, A can't be 1 or 3. So A must be 2.
Therefore, $A = 2$ and $B = 1$.

FOLLOW-UPS: (1) In the original problem, 6 squares were filled in to start you off. Can you fill in just four squares (one each of M, A, T, and H) so that all the other squares can be filled in uniquely according to the conditions of the problem? [Many possible answers exist. One is at the right.] *(2) How many different solutions exist?* [576: 24 ways to fill boxes × 24 ways to arrange the letters once the boxes are chosen.]

4E. *Strategy: Work backwards.*
Since we are working backwards, the entire table is reversed.

Coins in bank AFTER action	Action	Coins in bank BEFORE action
15 (Final result)	Took out 1/2 of the coins	30
30	Put in 4 coins	26
26	Took out 1/3 of the coins	39**
39	Put in 20 coins	19
19	Took out 1/2 of the coins	38 (Original amount)

To begin with, the bank contained 38 coins.

****Explanation of "39 coins":** After Sandy removed one-third of the coins, the 26 remaining coins were two-thirds of the amount before that removal. This number must be twice as large as the one third that she removed in this step. She must have removed 13 coins.

SET 6   **Olympiad 5**

5A. At the end of 3 days Bob has 50 stamps. At the end of 7 more days he has an additional $7 \times 4 = 28$ stamps. **Thus, Bob has a total of** $50 + 28 = $ **78 stamps at the end of 10 days.**

FOLLOW-UPS: (1) After how many days will he have 266 stamps? [57] (2) After how many days will he first have more than 500 stamps? [116]

5B. *Strategy: Consider different ways to get four identical digits.*
The odometer reading is greater than 62,222 and the leftmost digit remains 6. To get four identical digits, either the other four digits are identical to each other or three of them match the 6. The least number of the first type is 63,333. The least number of the second type is 62,666. **The least number of miles Mr. Jackson has to travel is** $62,666 - 62,222 = $ **444 miles.**

5C. *Strategy: Compare the sums in just two circles at a time.*
The sum of the numbers in circle A is the same as the sum of the numbers in circle C. Since Q and R are in both circles and 6 is 5 more than 1, S must be 5 more than P. Of the available numbers, the only difference of 5 is between 7 and 2. Thus $S = 7$ and $P = 2$. Next consider circle A and circle B. Circle A contains 6, 2, Q, and R. Circle B contains 4, 2, Q and 7. Q is in both circles, so $6 + 2 + R = 4 + 2 + 7$, and $R = 5$. Q is therefore 3, and **the sum of the numbers in each circle is 16.**

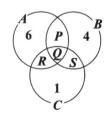

FOLLOW-UP: Start with the 3 circles as given in the problem, but with no numbers filled in. The whole numbers from 1 through 7 are placed, one in each region, so that the sum of the numbers in each circle is the same. What is the least possible value for this sum? [13] The greatest? [19]

5D. <u>**METHOD 1:**</u> *Strategy: Count in an organized way: exterior vs. interior squares.*

Each face of the figure has 8 painted squares for a total of $6 \times 8 = 48$ painted faces. Inside each "hole" 4 squares are painted for a total of $6 \times 4 = 24$ painted faces. **The total number of square units of painted surface is** $24 + 48 =$ **72.**

<u>**METHOD 2:**</u> *Strategy: Count in an organized way: corner vs. interior cubes.*

There are 8 corner cubes and $20 - 8 = 12$ non-corner cubes. Each corner cube has 3 painted faces for a total of $8 \times 3 = 24$ square units of painted surface. Each non-corner cube has 4 painted faces for a total of $12 \times 4 = 48$ square units of painted surface. The total number of painted faces is $24 + 48 = 72$.

<u>**METHOD 3:**</u> *Strategy: Start with a simpler problem.*

If there were no holes, each of the 6 faces of the cube would have 9 square units of painted surface for a total of 54 square units. For each center cube that is removed, we lose 1 square unit of painted surface but gain 4 square units of painted surface around the inside of the hole. This is a net gain of 3 square units of painted surface for each hole. Since there are 6 holes, the net gain is 6×3 or 18 square units, and the total painted surface is $54 + 18 = 72$ square units.

FOLLOW-UP (for the very adventurous): Explore this same situation with glued cubes measuring $4 \times 4 \times 4$ with a two-unit "hole" and $5 \times 5 \times 5$ with one- and/or three-unit "holes".

5E. <u>**METHOD 1A:**</u> *Strategy: Keep the sum constant and change the difference.*

Suppose the game is a tie and the sum of the scores is 44; then each team would have scored 22 points. In order to create a difference of 20 points, increase the Patriot score by 10 points and decrease the Giant score by 10 points, for a final score of 32-12. **The Patriots scored 32 points.**

<u>**METHOD 1B:**</u> *Strategy: Keep the difference constant and change the sum.*

Suppose one team scored the first 20 points, attaining the desired difference. The teams then split the remaining 24 points, 12 each. The final score is 32-12, with the Patriots scoring 32 points.

<u>**METHOD 2:**</u> *Strategy: Make a table, keeping the sum or difference constant.*

List several Patriot-Giant scores whose sum is 44, and look for a difference of 20. With a sum of 44 and a difference of 20, both scores are even.

Patriot score	Giant score	Sum	Difference
40	4	44	36
38	6	44	32
36	8	44	28
34	10	44	24
32	12	44	20

Alternately, a table that keeps the difference constant and changes the sums can be used.

METHOD 3:   *Strategy: Use algebra.*

Let G represent the Giants' score and $G + 20$ the Patriots' score. Now use the fact that the sum of their scores is 44 to set up an equation:

$$\underline{G} + \underline{G + 20} = 44 \quad [\text{"Giants' score + Patriots' score = 44 points."}]$$

Then: $2G + 20 = 44$ [Replace $G + G$ by $2G$ because they represent the same value.]

$2G = 24$ [Subtract 20 from each side of the equation.]

$G = 12$ [Divide each side of the equation by 2.]

$G + 20 = 32$ [The Patriots scored 20 points more than the Giants.]

The Patriots scored 32 points.

FOLLOW-UPS: Suppose the sum of the scores was 57. (1) Could the difference of the scores be 30? Explain. [No. 57 must be the sum of one odd and one even number. The difference would have to be odd.] *(2) How many different numbers could represent the difference?* [29]

SET 7	◆ ◆ ◆	**Olympiad 1**

1A. *Strategy: Rearrange the factors for easier multiplication.*
Write as $25 \times 4 \times 20 \times 17$.
Then $(25 \times 4) \times (20 \times 17) = 100 \times 340 = 34{,}000$.
The value of the product is 34,000.

1B. **METHOD 1:**   *Strategy: Start with a simpler problem.*
Remove the 2 people from the end seats. The remaining 20 people will be seated 2 to a table. This will require 10 tables. Reseating the 2 people in the end seats does not require any additional tables. Then **10 tables placed end to end will seat 22 people.**

METHOD 2:   *Strategy: Separate the tables into end tables and interior tables.*
Each end table holds 3 people and each interior table holds 2 people. The two end tables together seat 6 people. The interior tables hold the remaining 16 people, at 2 people to a table. Thus, eight interior tables are needed, for a total of ten tables.

METHOD 3:   *Strategy: Use simpler numbers to make a chart and then look for a pattern.*
Double the number of tables and then add 2 to arrive at the number of people that can be seated.

No. of tables	1	2	3	4	...	?
Number of people	4	6	8	10	...	22

Therefore, 10 tables are needed to seat 22 people because $(10 \times 2) + 2 = 22$.

FOLLOW-UPS: (1) If 15 four-cm squares are lined up end to end to form a rectangle, what will be the perimeter of the rectangle? [128 cm] *(2) If the tables seat 3 people on each side, how many tables must be placed end to end to seat 100 people?* [16 tables, with 2 empty seats]

1C. **METHOD 1:** *Strategy: Decide how adding 24 girls changes the ratio.*

The number of boys stays the same. However, the addition of 24 girls changes the number of girls for every boy from 2 to 5. That is, a total of 24 more girls results in 3 more girls for every boy. Since $24 \div 3 = 8$, **8 boys are in the group.**

METHOD 2: *Strategy: Choose simpler numbers to make a table.*

Number of Boys	1	2	3	...	?
Number of Girls (*before*) [B]	2	4	6	...	
Number of Girls (*After*) [A]	5	10	15	...	
Number of Girls Joining [A – B]	3	6	9	...	24

The first row shows there are 1, 2, 3, ... boys. The second row shows the initial 2 girls for every boy and the third row shows the subsequent 5 girls for each boy. The fourth row shows the difference between rows 2 and 3; that is, the number of girls who joined later. Compare the fourth row to the first row. The number of boys is one-third of the number of girls who joined the group. Since 24 girls actually joined, there must be 8 boys.

1D. **METHOD 1:** *Strategy: Find the multiples of exactly two of the three time intervals.*

Consider each pair of owls in turn. Let owl *A* hoot every 3 hours, owl *B* hoot every 8 hours, and owl *C* hoot every 12 hours.

1. Owls *A* and *B* hoot together at any hour that is a multiple of both 3 and 8. That is a multiple of 24, and therefore also a multiple of 12. Whenever owls *A* and *B* both hoot, owl *C* hoots too.

2. Owls *B* and *C* hoot together at any hour that is a multiple of 8 and 12. That is a multiple of 24, and 24 is also a multiple of 3. Whenever owls *B* and *C* both hoot, owl *A* hoots too.

3. Owls A and C hoot together at any hour that is a multiple of 3 and 12. That is a multiple of 12. The multiples of 12 that are within the required time are 12, 24, 36, 48, 60, and 72. Of these only 12, 36, and 60 are not multiples of 8. **Exactly two owls hoot together a total of 3 times.**

METHOD 2: *Strategy: List the multiples, largest first. Which multiples occur twice?*

Multiples of 12		12	24	36	48	60	72
Multiples of 8	8	16	24	32	40	48	56	64	72
Multiples of 3	3 6 9	12	15 18 21 24 27 30 33	36	39 42 45 48 51 54 57	60	63 66 69 72 75 78

FOLLOW-UPS: *(1) What is the first hour after the 1000th hour when exactly two owls hoot together?* [1020] *(2) How many times does one owl hoot alone during the first 60 hours?* [20]

SOLUTIONS: DIVISION E

1E. _Strategy: Draw a picture of each of the two possibilities._
Either Dawn is at Amy's right (_figure 1_) or left (_figure 2_).

Suppose _figure 1_ shows the correct position of the girls. The girl to Amy's right (Dawn) has the green ring. The girl to Soumiya's left (also Dawn) has the red ring. Then Dawn has two rings, which contradicts the statement in the problem.

So _figure 2_ shows the correct position. The girl at Amy's right, with the green ring, is Soumiya. The girl at Soumiya's left, Amy, has the red ring. Dawn has the remaining color, blue. **Amy's ring is red. Dawn's ring is blue. Soumiya's ring is green.**

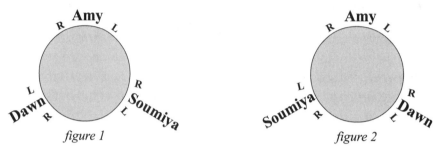

figure 1   figure 2

R = her right, L = her left

SET 7   **Olympiad 2**

2A. **METHOD 1:** _Strategy: Work from the largest container inward._
Find how many of each item one case can hold.
Each case holds 2 cartons.
Each carton holds 3 boxes, so one case holds 2 × 3 = 6 boxes.
Each box holds 4 bundles, so one case holds 6 × 4 = 24 bundles.
Each bundle holds 5 envelopes, so one case holds 24 × 5 = 120 envelopes.
Each envelope holds 6 pencils, so **one case holds** 120 × 6 = **720 pencils.**

METHOD 2: _Strategy: Work from the smallest container outward._
Find how many pencils each item can hold.
Each envelope holds 6 pencils.
Each bundle holds 5 envelopes, so one bundle holds 5 × 6 = 30 pencils.
Each box holds 4 bundles, so one box holds 4 × 30 = 120 pencils.
Each carton holds 3 boxes, so one carton holds 3 × 120 = 360 pencils.
Each case holds 2 cartons, so one case holds 2 × 360 = 720 pencils.

2B. _Strategy: First determine how much oil is on hand._

If 4 barrels are used each day and the supply lasts 30 days, the amount available must be $30 \times 4 = 120$ barrels. Therefore, to last 40 days, $120 \div 40 =$ **3 barrels should be used per day.**

FOLLOW-UPS: _(1) Suppose 12 carpenters can build a house in 42 days. In how many days can 14 carpenters each working at the same pace build an identical house?_ [36] _(2) A group of six scouts purchase rations sufficient for a 15-day camping trip. If 3 more scouts join the group but no more rations are purchased, how many days will the rations last?_ [10 days]

2C. _Strategy: Focus on the multiplicand._

A is 1, since any greater value produces a 5-digit product when multiplied by 9.

$$\begin{array}{r} 1\,B\,C\,D \\ \times\qquad 9 \\ \hline D\,C\,B\,1 \end{array}$$

D is 9, since only 9 times 9 results in a 1 in the ones place.

$$\begin{array}{r} 1\,B\,C\,9 \\ \times\qquad 9 \\ \hline 9\,C\,B\,1 \end{array}$$

B is 0, since 9 times any other number (including 1) results in a "carry" into the thousands place.

$$\begin{array}{r} 1\,0\,C\,9 \\ \times\qquad 9 \\ \hline 9\,C\,0\,1 \end{array}$$

C is 8, since 9 times C ends in 2 so that when the 8 "carry" is added, 0 is produced in the tens place.

$$\begin{array}{r} 1\,0\,8\,9 \\ \times\qquad 9 \\ \hline 9\,8\,0\,1 \end{array}$$

The four-digit number _ABCD_ is 1089.

FOLLOW-UPS: _(1) If AB × 6 = BBB, what is the value of BBB?_ [444] _(2) The square of what number is 1089?_ [33]

2D. _Strategy: Find the least and greatest multiples of 7 between 100 and 1000._

$100 \div 7 = 14\frac{2}{7}$, so consider $14 \times 7 = 98$ or $15 \times 7 = 105$.

$1000 \div 7 = 142\frac{6}{7}$, so consider $142 \times 7 = 994$.

The problem then is to consider the number of multiples of 7 from 98 (or 105) to 994.

METHOD 1: _Strategy: Consider the number of multiples less than 100 and 1000._
There are 142 multiples of 7 less than 1000. Of these, the first 14 should not be counted because they are less than 100. The number that should be counted is $142 - 14 = 128$. **There are 128 multiples of 7 between 100 and 1000.**

METHOD 2: _Strategy: Consider the greatest multiples less than 100 and 1000._
The difference between the greatest multiples of 7 less than 1000 and less than 100 is $994 - 98 = 896$. There are $896 \div 7 = 128$ multiples of 7.

METHOD 3   *Strategy: Consider the number of multiples between 100 and 1000 inclusive.*
Suppose all the multiples of 7 that are less than 1000 are listed in order: 7, 14, ..., 98, ..., 994.
Then the least and greatest multiples of 7 between 100 and 1000 are 105 and 994, which are the
15th and 142nd numbers on the list, respectively. The difference in the positions is $142 - 15 = 127$.
But this does not include the 105. In all there are 128 multiples of 7.

FOLLOW-UPS: *(1) How many multiples of 6 are between 100 and 1000?* [150] *(2) How many of these are
also multiples of 7?* [21]

2E. **METHOD 1:**   *Strategy: Find areas of the whole figure and the garden alone. Then subtract.*
The sidewalk extends the dimensions of the garden as shown.

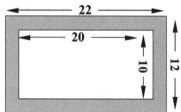

Area of garden and sidewalk together: 22 m × 12 m   =   264 m²
Area of garden alone: 20 m × 10 m   =   200 m²
                                 The area of sidewalk is 64 m².

METHOD 2:   *Strategy: Partition the sidewalk into pieces whose area can be found. Then add.*
Each of the 4 pieces labeled *A* has area $1 \times 1 = 1$ m².
Each of the 2 pieces labeled *B* has area $20 \times 1 = 20$ m².
Each of the 2 pieces labeled *C* has area $10 \times 1 = 10$ m².
The total area of the sidewalk is $(4 \times 1) + (2 \times 20) + (2 \times 10) =$
$4 + 40 + 20 = 64$ m².

FOLLOW-UP: *A rectangular patio 12 m by 20 m is covered with square 2-meter tiles. The brown tiles form
the border, and the tan tiles form the interior. How many tan tiles are there?* [32]

3A. *Strategy: Proceed one condition at a time. Start with the most limiting condition.*
  • Multiples of 12: 12, 24, 36, 48, 60, 72, 84, 96
  • The sum of the digits is greater than 10: 48, 84, 96
  • The difference between the digits is greater than 3: 48, 84
  • The tens digit is larger than the ones digit: 84
The number is 84.

SOLUTIONS: DIVISION E

3B. **METHOD 1:**   *Strategy: Consider one place at a time.*

Any one of the 3 available numerals can be chosen for the first digit. For each of these 3 choices any of the 3 numerals can be chosen for the second digit, so there are $3 \times 3 = 9$ ways to choose the first two digits. Again, there are 3 choices for the last digit for each of the 9 possibilities. Therefore, 9×3 or **27 different three-digit house numbers can be made.**

METHOD 2:   *Strategy: List the numbers in an organized way.*

For example,

444	474	484	744	774	784	844	874	884
447	477	487	747	777	787	847	877	887
448	478	488	748	778	788	848	878	888

These are the 27 numbers.

This method can be made visual by using a tree diagram.

FOLLOW-UPS: (1) How many different five-digit house numbers could be made using just 4, 7, and 8? [243] (2) In the original problem, what is the probability that an address chosen at random has exactly 2 digits the same? [$\frac{18}{27}$ or $\frac{2}{3}$]

3C. *Strategy: Look for a pattern to make the computation easier.*

Notice that the problem lists all the even numbers from 100 to 2. The complete expression has 50 numbers — group them into 25 pairs: $(100 - 98) + (96 - 94) + (92 - 90) + \ldots + (4 - 2)$. The value of each pair in parentheses is 2. **The value is** $25 \times 2 = $ **50.**

FOLLOW-UP: Find the sum of all the whole numbers from 1 through 100. [5050]

3D. *Strategy: Draw diagrams. Try simpler related cases.*

If **one straight line** is drawn, the interior of the circle can be cut into a maximum of 2 regions.

If **two straight lines** are drawn, the interior can be cut into a maximum of 4 regions. Note that this requires the 2 lines to intersect. If they don't intersect, only 3 regions are produced.

If **three straight lines** are drawn to produce the maximum number of regions, the third line must intersect *each* of the other lines at *distinct* points, as shown at the far right.

The maximum number of regions is 7.

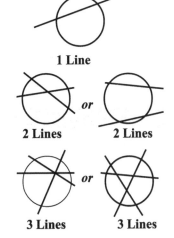

1 Line

2 Lines   *or*   2 Lines

3 Lines   *or*   3 Lines

FOLLOW-UPS: (1) What is the greatest number of regions into which 4 lines can cut a circle? 5 lines? [11,16] (2) Compare these results with the first few triangular numbers. [Each is one more than the corresponding triangular number 1, 3, 6, 10, ...] (3) Do the results change if the circle is replaced by a rectangle? [No]

3E. _Strategy: Move the people in the three rows being eliminated._

3 fewer than 18 rows is 15 rows, each of which seats 6 more people. Those 15 rows seat an additional $15 \times 6 = 90$ people. Those 90 people came from the 3 rows that were removed, so each original row held 30 people. **There are** $30 \times 18 = $ **540 seats in the theater.**

Check your answer by checking the new configuration of the theater: 15 rows now seat 36 people each, also for a total of 540 people.

SET 7	Olympiad 4

4A. _Strategy: Compare Meg's age to Beth's two different ways._

Jo is 3 years older than Beth and Meg is 4 years older than Jo, so Meg is 7 years older than Beth. Meg is also twice as old as Beth. The only way that adding 7 to Beth's age could result in a number twice as large is if Beth is 7. Then Meg is 14, Jo is 10, and **Amy is 5.**

4B. _Strategy: Use the nature of place values to maximize the sum._

The greatest sum occurs when the left-most (thousands) digits are maximized. Then A and C are 2 and 3 (in some order), making $B = 1$. The next most important digits in determining the sum are in the second (hundreds) column. Since $B = 1$, this sum will be greatest when $A = 3$. Then C must be 2. **The greatest sum is** $3124 + 5312 + 2163 = $ **10,599.**

Follow-Up: Suppose that A, B, C, and D represent 1, 2, 3, and 4 in some order. What is the least value of the sum? [7317]

$$\begin{array}{r} A\ B\ C\ 5 \\ B\ D\ 6\ C \\ +\ C\ 7\ A\ 0 \\ \hline \end{array}$$

4C. _Strategy: Choose the gumballs so as to avoid three of a color._

After Anna buys 8 gumballs, she might have 2 red, 2 green, 2 yellow, and 2 purple. The next gumball she buys must be one of those same four colors, guaranteeing her three of the same color. So, the minimum number of gumballs she should expect to buy is 9. **The minimum cost that guarantees three gumballs of the same color is** $5 \times 9 = $ **45¢.**

Follow-Ups: The machine has 15 red, 20 green, 25 yellow, and 30 purple gumballs. (1) At 5¢ each, how much must she spend to guarantee that she has 3 yellow gumballs? [$3.40] _(2) For which color does it cost the most to guarantee that all three gumballs are that color? How much?_ [Red; $3.90]

4D. _Strategy: Use factors._

Because the area of a rectangle is length × width, the length and width must be a factor pair of 36. Take each factor pair of 36 as a possible length and width and determine the perimeter in each case.

Width	Length	Perimeter	
1	36	1 + 36 + 1 + 36 =	74
2	18	2 + 18 + 2 + 18 =	40
3	12	3 + 12 + 3 + 12 =	30
4	9	4 + 9 + 4 + 9 =	26
6	6	6 + 6 + 6 + 6 =	24

The greatest possible perimeter is 74 cm.

FOLLOW-UPS: _(1) A rectangle has an area of 360 sq cm. The length and width are whole numbers. What is the least perimeter possible?_ [76 cm] _(2) State a general rule for this kind of problem._ [If the area of a rectangle is known, the perimeter is least when the length and width are most nearly equal (when the rectangle is a square or as close as possible). The perimeter is greatest when the length and width are farthest apart.]

4E. **METHOD 1A:** _Strategy: Start with an extreme case — all motorcycles._

Suppose all 70 vehicles were motorcycles. At 2 tires each the motorcycles would require 140 tires. This is 269 – 140 or 129 tires less than the company produced. Each car needs 3 more tires than a motorcycle. Then 129 ÷ 3 = 43 of the motorcycles must be "traded in" for 43 cars to bring the total number of tires up to 269. Therefore 70 – 43 = 27 of the vehicles were motorcycles. To check: (43 × 5) + (27 × 2) = 215 + 54 = 269 tires. **The company produced 54 motorcycle tires.**

METHOD 1B: _Strategy: Start with an extreme case — all cars._

Suppose all 70 vehicles were cars. At 5 tires each the cars would require 350 tires. This is 350 – 269 = 81 tires more than the company produced. Each motorcycle needs 3 fewer tires than a car. Then 81 ÷ 3 = 27 cars must be "traded in" for 27 motorcycles to bring the total number of tires down to 269. The 27 motorcycles require 27 × 2 = 54 tires.

METHOD 2: _Strategy: Assume half the vehicles are motorcycles._

Suppose there are 35 cars and 35 motorcycles, then there are (35 × 5) + (35 × 2) = 245 tires. This is 24 tires short of the stated total of 269. Since each car has 3 more tires than each motorcycle, there are 24 ÷ 3 = 8 more cars and 8 fewer motorcycles. Because there are 27 motorcycles, there are 54 motorcycle tires.

SOLUTIONS: DIVISION E

5A. <u>**METHOD 1:**</u>   *Strategy: Work backwards, using inverse (opposite) operations.*

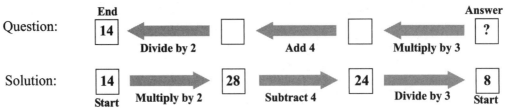

Toni started with 8.

<u>**METHOD 2:**</u>   *Strategy: Use algebra.*

Let N = the number Toni started with.

Then $\frac{3N + 4}{2} = 14$

Multiply both sides by 2: $3N + 4 = 28$

Subtract 4 from each side: $3N \quad = 24$

Divide each side by 3: $N \quad = \; 8$

Thus, Toni started with 8.

5B. <u>**METHOD 1:**</u>   *Strategy: Use a date line to name each "event".*

Place each event on the line, starting with "5 days before..." Label it Friday.

Therefore, **yesterday was Sunday.**

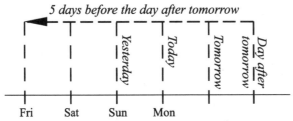

<u>**METHOD 2:**</u>   *Strategy: Count the days from today.*

The day after tomorrow is 2 days after today.

Five days before the day after tomorrow is 3 days before today.

If 3 days ago was Friday, today is Monday, and yesterday was Sunday.

5C. _Strategy: Relate smiles and laughs by making the number of grins the same in each relation._
METHOD 1: _Strategy: Simplify 6 grins = 9 laughs._
If 6 grins = 9 laughs, then for every 2 grins, there are 3 laughs. Rewrite as 10 grins = 15 laughs.
But 10 grins also equals 3 smiles, so that 3 smiles = 15 laughs. Then 1 smile = 5 laughs and **it takes 2 smiles to equal 10 laughs.**

METHOD 2: _Strategy: Combine both relations using the LCM of the number of grins._
The least common multiple of 6 grins and 10 grins is 30 grins.
Since 6 grins = 9 laughs, then 30 grins = 45 laughs.
Since 10 grins = 3 smiles, then 30 grins = 9 smiles.
Therefore 9 smiles= 45 laughs.
Then for every smile there must be 5 laughs.
Thus it takes 2 smiles to equal 10 laughs.

5D. _Strategy: Find the dimensions of the box. Then add the areas of the faces._
Denote the dimensions of the box by _a_, _b_, and _c_.
The base is a square of area 16 sq in, so _a_ and _b_ are each 4 in.
The given volume of 80 cu in = $a \times b \times c = 4 \times 4 \times c$, so _c_ is 5 in.
The top and bottom each have an area of 16 sq in.
Each side has an area of $4 \times 5 = 20$ sq in.
The amount of paper needed is $16 + 16 + 20 + 20 + 20 + 20 = $ **112 sq in.**

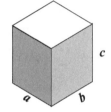

FOLLOW-UP: Warren's box has a square base of area 36 sq cm. He needs 168 sq cm of paper to completely cover the box, including the top and bottom. What is the volume of Warren's box? [144 cu in]

5E. _Strategy: Use the divisibility tests for 5, 4, and 3 to determine each digit._
1. _BCD_ is divisible by 5, so _D_ = 5. That leaves 1, 2, 3, and 4 for _A_, _B_, _C_, and _E_.
2. _ABC_ is divisible by 4 only if _BC_ is divisible by 4. _BC_ = 12 or 32 or 24.
3. _CDE_ is divisible by 3 only if _C_ + _D_ + _E_ is divisible by 3. Then _C_ + 5 + _E_ is divisible by 3, so _C_ + _E_ is 4 or 7.
4. Suppose _C_ + _E_ = 4. We know from step 2 that _C_ is 2 or 4. If _C_ = 2, then _E_ is also 2. This is impossible because each digit must be used exactly once. If _C_ = 4, then _E_ = 0, which is not one of the given values. So _C_ + _E_ = 7.
5. If _C_ = 2, then _E_ would equal 5 and have the same value as _D_. So _C_ = 4. Then _E_ = 3, and from step 2, _B_ = 2. Finally, _A_ must be 1, and **the five-digit number is 12,453.**

FOLLOW-UPS: Investigate other divisibility tests. For what values of A is the six-digit number 3A9308 divisible by 3? [1 or 4 or 7] _by 4?_ [any] _by 6?_ [1 or 4 or 7] _by 7?_ [2 or 9] _by 8?_ [none] _by 9?_ [4] _by 11?_ [1] _by 13?_ [9]

1A. **METHOD 1:** *Strategy: Use the Associative Principle for Multiplication.*
$42 \times 25 = (21 \times 2) \times 25$.
This can be written as $21 \times (2 \times 25) = 21 \times 50$.
Thus $42 \times 25 = 21 \times 50$.
The number is 50.

METHOD 2: *Strategy: Perform the indicated operations.*
$42 \times 25 = 1,050$. Then $1,050 \div 21 = 50$. The number is 50.

FOLLOW-UPS: *(1) $24 \times 27 = 4 \times \underline{?}$. [162] (2) $36 \times 75 = 300 \times \underline{?}$. [9] (3) Use the Associative Principle for Multiplication to multiply 125 by 48 mentally. [6000]*

1B. *Strategy: Apply one condition at a time.*
The hundreds digit equals the sum of the other two digits. That means that the hundreds digit is half the sum of all three digits. The hundreds digit is 6. The hundreds digit is one greater than the tens digit, so the tens digit is 5. Finally, the sum of the three digits is 12. $6 + 5 + 1 = 12$, so the ones digit is 1. **The number is 651.**

1C. **METHOD 1:** *Strategy: Use their differences in expenditures.*
For every $3 Marisa spends, Andie spends $5, a difference of $2. To get a difference of $120, each person would have to spend $120 \div $2 = 60$ times as much. Then Marisa spends $60 \times $3 = 180 and **Andie spends** $60 \times $5 = 300. Check: The difference is $300 - 180$ or $120.

METHOD 2: *Strategy: Make a chart and look for a difference of $120.*

Amount Andie Spent	$50	$100	$150	$200	$250	**$300**
Amount Marisa Spent	$30	$60	$90	$120	$150	**$180**
Difference	$20	$40	$60	$80	$100	**$120**

1D. *Strategy: Subtract areas.*
The area of a rectangle is obtained by multiplying the length by the width. If the rectangle measures 4 cm by 12 cm, the total area is 48 cm². The sum of the areas of the other 3 triangles is $16 + 18 + 8$ or 42 cm². **The area of the shaded triangle is $48 - 42$ or 6 cm².**

FOLLOW-UPS: *(1) An interior point of a rectangle is joined to the four vertices forming four triangles. Suppose the rectangle is 6 cm by 10 cm and the areas of the four triangles are consecutive even integers. What is the area of the smallest triangle? [12 cm²] (2) Suppose the rectangle is 7 m by 12 m. What is the sum of the areas of the top and bottom triangles? [42 m²] (3) Can you explain why the sum of these two areas is always half the area of the rectangle?*

SOLUTIONS: DIVISION E

1E. **METHOD 1:** *Strategy: Use the least common multiple (LCM) of 6 and 9.*
The number of coins is a multiple of both 6 and 9. It must be a multiple of their LCM, 18. The number of coins is also 2 more than a multiple of 5. Then the number of coins ends in 2 or 7. No multiple of 18 ends in 7. The least multiple of 18 that ends in 2 is $4 \times 18 = 72$. **The least number of coins Jeff could have is 72.**

METHOD 2: *Strategy: List numbers satisfying one condition. Find which satisfy the others.*
If Jeff has 2 coins remaining when he puts the same number of coins in each of 5 bags, his total number of coins is 2 more than a multiple of 5. Those possibilities are: 7, 12, 17, 22, 27, 32, 37, 42, 47, 52, 57, 62, 67, 72, ... Of these, the multiples of 6 are 12, 42, 72, ... The least multiple of 9 on the second list is 72. The least number that satisfies all 3 conditions is 72.

FOLLOW-UPS: *(1) What is the next three least number of coins after 72 that Jeff could have?* [162, 252, 342]. *(2) Find the pattern and explain why it works.* [The numbers increase by 90, which is the LCM of 5, 6, and 9.]

2A. **METHOD 1:** *Strategy: Count from the average (mean).*
If the sum of five numbers is 45, their average (mean) is $45 \div 5$ or 9. Because the numbers are consecutive (and because there is an odd number of them), the middle number is 9. Counting backwards, the second number is 8, and **the least of the five numbers is 7.**

METHOD 2: *Strategy: Start with any 5 consecutive numbers. Then adjust the sum.*
Start with 1, 2, 3, 4, 5, for example. $1 + 2 + 3 + 4 + 5 = 15$. This is 30 less than the given sum of 45. To keep the numbers consecutive but increase the total by 30, each number must be increased by $30 \div 5 = 6$. The numbers are $1 + 6$, $2 + 6$, $3 + 6$, $4 + 6$, and $5 + 6$; that is 7, 8, 9, 10, and 11. The least of the five numbers is 7.

FOLLOW-UPS: *(1) To receive an "A" for a course, Jennie needs at least an average (mean) of 90 on five exams. Jennie's grades on the first four exams are 84, 95, 86 and 94. What is the minimum score she needs on the fifth exam to receive an "A" for the course?* [91] *(2) (Exploration) What is the greatest number that must be a divisor for every sum of 5 consecutive whole numbers? 7? 9? 199?* [5, 7, 9, 199] *Can you reach a conclusion for the sum of an even number of consecutive whole numbers?*

SOLUTIONS: DIVISION E

2B. <u>METHOD 1:</u> *Strategy: Count the number of ways to place blue and green books only.*
Draw four spaces. Then place the blue and green books. The possible results are:

| B G □ □ | or | □ B G □ | or | □ □ B G | or |
| B □ G □ | or | □ B □ G | or | B □ □ G | |

In each case there are two spaces left for the red and yellow books. The red book must go in the left space and the yellow book in the right space. This does not change the number of possibilities listed above. **There are 6 different orders in which Sara can place the books.**

<u>METHOD 2:</u> *Strategy: List all possible orders. Then cross out all sets in the wrong order.*
Fill in the spaces right-to-left: For each of the 4 books, any of the 3 remaining books can occupy the space immediately to its left. For each of these $3 \times 4 = 12$ arrangements, either of 2 remaining books can occupy the third position, a total of $12 \times 2 = 24$ arrangements. With only 1 book left for the one position, there are 24 possible orders for the four books. In half of them the blue and green books are in the wrong order. In half of the remaining 12 orders, the red and yellow books are in the wrong order. In only the remaining 6 orders (listed in Method 1) are all the books correctly arranged.

FOLLOW-UP: Suppose Sara also has a white book, making five books in all. How many orders now are possible? [30]

2C. *Strategy: Find the number of tablets in one bottle.*
Ann, Bill and Chris each take 2 tablets daily, and Wendy takes one daily. They take a total of 7 tablets each day. The total number of tablets in the bottle is exactly enough for 24 days. There are $24 \times 7 = 168$ tablets in one bottle. If Wendy takes an additional tablet each day, then they take 8 tablets per day. **The tablets in one bottle will last** $168 \div 8 = \textbf{21 days.}$

FOLLOW-UP: A group of 10 Girl Scouts purchases food for a week-long outdoor trip. If four more girls join the group, but no more food is purchased, how many days will the food last? [5]

2D. *Strategy: Find the possible lengths of the sides.*
The perimeter, 14 m, is the sum of two lengths and two widths. Then the sum of one length and one width (called the semiperimeter) is 7 m. The possible dimensions are:
- 6 m by 1 m, giving an area of 6×1 or 6 m²
- 5 m by 2 m, giving an area of 5×2 or 10 m²
- 4 m by 3 m, giving an area of 4×3 or 12 m² .

The area could be 6, 10 or 12 m².

SOLUTIONS: DIVISION E

2E. **METHOD 1:** *Strategy: Count the number of 4s appearing in the ones and in the tens place.*
There are $200 \div 10 = 20$ decades (1-10, 11-20, etc.). In each decade, the digit 4 appears in the ones place 1 time. In all it appears in the ones place 20 times.

There are two centuries (1-100, 101-200). In each century, the digit 4 appears in the tens place 10 times. In all it appears in the tens place 20 times. **The digit 4 will appear** a total of $20 + 20 = $ **40 times on Aiah's list.**

METHOD 2: *Strategy: Count the number of 4s appearing in each interval of 10 numbers.*

Interval	1-9	10-19	20-29	30-39	40-49	50-59	60-69	70-79	80-89	90-99
Number of 4's	1	1	1	1	11*	1	1	1	1	1

Note that 4 appears twice in the number 44.

The total number of 4s from 1 through 100 (100 has no 4s) is 20. In listing the numbers from 101 through 199, the ones and tens digits from the above list are repeated. This gives an additional 20 digits of 4. 200 contains no 4s. The digit 4 will appear 40 times on Aiah's list.

3A. **METHOD 1A:** *Strategy: Consider how adding 14 changes 4 times the starting number.*
Adding 14 changes the value from 4 times the starting number to 6 times the starting number. Then 14 must be 2 times the starting number. **The starting number is 7.**

METHOD 1B: *Strategy: Write an equation.*
Let $N = $ the starting number.   (Note: $4N$ means $4 \times N$)

$$4N + 14 = 6N \qquad \text{(Now subtract } 4N \text{ from each side)}$$
$$4N - 4N + 14 = 6N - 4N \qquad \text{(Combine like terms)}$$
$$14 = 2N \qquad \text{(Divide each side by 2)}$$
$$7 = N \qquad \text{(Thus the starting number is 7)}$$

The algebraic procedures in Method 1B is the same as the reasoning in Method 1A.

METHOD 2: *Strategy: Make a table. Find a pattern.*

Starting number (SN)	1	2	3	...	?
SN × 4	4	8	12	...	
SN × 4 + 14	18	22	26	...	
SN × 6	6	12	18	...	
Row 3 – row 4	12	10	8	...	0

Notice that as the starting number increases by 1, the difference between 6 times the starting number and (4 times the starting number + 14) decreases by 2. These two will be equal when the difference is 0. That takes 4 more steps. The starting number is 7. Check that $7 \times 4 + 14 = 6 \times 7$.

3B. *Strategy: Find out how many games Hannah played.*
4 minutes equals 8 sets of 30 seconds, so Hannah's 4-minute game was her eighth game. Hannah played 8 games, spending 25¢ each. **Hannah has spent $2.00.**

FOLLOW-UP: What was the total time Hannah spent playing the video game? [18 minutes]

3C. *Strategy: Compare Dwayne's number to the average.*
There are 12 houses and 57 letters. In order for more letters to be delivered to Dwayne than to anyone else and yet be a minimum, the letters must be distributed as equally as possible. Delivering 4 letters to each house leaves 9 letters undelivered. Since more letters are delivered to Dwayne's house than any other, 2 of the 9 letters are delivered to him and 1 of them to each of seven other houses. Thus 6 letters go to his house, 5 letters go to each of the seven houses and and 4 letters go to each of the remaining four houses. (If Dwayne were to get 5 letters and each of the other houses 4 or fewer letters, the greatest number of letters possible would be 49.) **The least number of letters that Dwayne could have received today is 6.**

3D. *Strategy: Use common factors to determine side-lengths.*
Side $\overline{HJ}$ is common to rectangles $ABJH$ and $HJFG$. Its length is a common factor of 6 and 9: 1 cm or 3 cm. It cannot be 1 or else BJ would be 6, which is not a factor of 10. Thus $HJ = 3$.

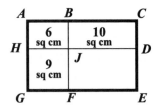

Similarly, $\overline{BJ}$ is common to rectangles $ABJH$ and $BCDJ$ and its length is a common factor of 6 and 10. Thus, by the same reasoning, $BJ = 2$.

METHOD 1: *Strategy: Combine the areas of all four small rectangles.*
In rectangle $BCDJ$, since $BJ = 2$, $JD = 5$.
In rectangle $HJFG$, since $HJ = 3$, $JF = 3$.
As a result, the area of rectangle $JDEF$ is $5 \times 3 = 15$ sq cm.
Therefore, **the total area of rectangle $ACEG$ is** $6 + 10 + 9 + 15 = $ **40 sq cm.**

METHOD 2: *Strategy: Multiply the sides of the large rectangle.*
In rectangle $ABJH$, since $HJ = 3$, $AB = 3$.
In rectangle $BCDJ$, since $BJ = 2$, $BC = 5$.
Thus, $AC = 8$.

In rectangle $ABJH$, since $BJ = 2$, $AH = 2$.
In rectangle $HJFG$, since $HJ = 3$, $HG = 3$.
Thus, $AG = 5$.

Therefore, the area of rectangle $ACEG$ is $8 \times 5 = 40$ sq cm.

SOLUTIONS: DIVISION E

3E. _Strategy: Make an organized list._

Only with the total score matters, not the order in which the darts are thrown. Find all possible scores starting with 0 and work upwards until the least total score that is impossible is reached. Can you find other ways to reach some of these totals?

$0 = 0 + 0 + 0$	$5 = 3 + 1 + 1$	$10 = 8 + 1 + 1$
$1 = 1 + 0 + 0$	$6 = 3 + 3 + 0$	$11 = 8 + 3 + 0$
$2 = 1 + 1 + 0$	$7 = 3 + 3 + 1$	$12 = 8 + 3 + 1$
$3 = 3 + 0 + 0$	$8 = 8 + 0 + 0$	$13 = $ impossible
$4 = 3 + 1 + 0$	$9 = 8 + 1 + 0$	

The least total score that is impossible to obtain is 13.

FOLLOW-UPS: _(1) The greatest possible score for 3 darts using this target is $15 + 15 + 15$ or 45. The least possible score is $0 + 0 + 0$ or 0. How many scores between 0 and 45 are impossible? [18] (2) If a 6-region dartboard has point-scores of 1 and 3 points in 2 of its regions, what point-scores are needed in the other 4 regions in order to make all scores from 1 to 25 possible using only 3 darts? [8, 13, 18, 23]_

SET 8   Olympiad 4

4A. **METHOD 1:**   _Strategy: Start with all 10¢ stamps and adjust._

Suppose Madison buys one dollar's worth of 10¢ stamps. She has ten of them. Each 10¢ stamp can be exchanged for two 5¢ stamps and each exchange increases the total number of stamps by one. To get 13 stamps, trade three 10¢ stamps for six 5¢ stamps. This leaves seven 10¢ stamps. **Madison gets six 5¢ stamps.** To check, $6 \times 5¢ + 7 \times 10¢ = \$1.00$.

(Note: the same approach can be used by starting with a dollar's worth of five-cent stamps, or with 13 ten-cent stamps, or with 13 five-cent stamps.)

METHOD 2:   _Strategy: Make a list and look for a pattern._

Number of 10¢ stamps	0	1	2	...	?
Number of 5¢ stamps	13	12	11	...	?
Total value	65¢	70¢	75¢	...	$1.00

In each column, the total number of stamps is 13. Notice that each time a 5¢ stamp is exchanged for a 10¢ stamp, the total value of the stamps increases by 5 cents. Starting with 65¢, the value must be increased by 35¢ to get \$1.00. This means that seven 5¢ stamps must be exchanged for seven 10¢ stamps, leaving 6 five-cent stamps.

4B. **METHOD 1:**   _Strategy: Consider the ones digit first._

The number is even, so the ones digit is even: 0, 2, 4, 6, and 8 are the five possibilities. The sum of the digits is odd, so the tens digit is odd: 1, 3, 5, 7, and 9 are the five possibilities. For each possible ones digit, there are five possible tens digits. Altogether, **there are 5×5 or 25 even two-digit numbers that have an odd number as the sum of their digits.**

Solutions - Division E Olympiads   215

SOLUTIONS: DIVISION E

METHOD 2: *Strategy:* *Make an organized list.*

10	12	14	16	18
30	32	34	36	38
50	52	54	56	58
70	72	74	76	78
90	92	94	96	98

There are 25 numbers.

4C. METHOD 1: *Strategy: Convert the sequence into a more familiar sequence.*
The numbers in my list are 3, 7, 11, 15, …, 99. If I add 1 to each number, my list becomes 4, 8, 12, 16, …, 100. These are the multiples of 4. Since $100 \div 4 = 25$, 100 is the 25th number on the new list, so 99 is the 25th number on the original list. **The value of N is 25.**

METHOD 2: *Strategy: Use the fact that each number is 4 more than the one before it.*
The first number is 3. The second number is $3 + 4$. The third number is $3 + 4 + 4$, and so on. The interval from the first to the 25th number is $99 - 3 = 96$, and $96 \div 4 = 24$. Therefore, 99 is twenty-four numbers after 3 in my list and so it is the twenty fifth number. The value of N is 25.

METHOD 3: *Strategy: Make a list.*
3, 7, 11, 15, 19, 23, 27, 31, 35, 39, 43, 47, 51, 55, 59, 63, 67, 71, 75, 79, 83, 87, 91, 95, 99. There are 25 numbers on the list, so $N = 25$.

FOLLOW-UPS: *(1) If I continue my list, what will be the 1000th number?* [3999] *(2) If I start with 8 and count by 5's, which number on the list is 403?* [80th]

4D. METHOD 1: *Strategy: Find the area of each face of the entire solid.*
The front face of the entire solid is a rectangle with length 30 cm and width 1 cm. Its area is 30×1 or 30 cm². The top, back and bottom faces are congruent to the front face. Each has an area of 30 cm². The left and right ends of the solid are squares 1 cm by 1 cm. Each has an area of 1 cm². **The surface area of the solid is $4 \times 30 + 2 \times 1 = $ 122 cm².**

METHOD 2: *Strategy: Find the exposed surface area of each of the cubes.*
Each interior cube, numbers 2 through 29, has 4 faces exposed. Each face has an area of 1 cm², so each of these cubes contributes 4 cm² to the surface area of the large solid. Both end cubes, numbers 1 and 30, have 5 faces exposed, so each of these cubes contributes 5 cm² to the surface area of the solid. The total surface area of the solid is $28 \times 4 + 2 \times 5$ or 122 cm².

FOLLOW-UPS: *(1) How should the 30 cubes be placed to form a solid one-layer high with the least possible surface area?* [Place them in 5 rows of 6 cubes each (or 6 rows of 5 cubes each) for a total surface area of 82 cm².] *(2) How should the cubes be placed to form a solid of more than one layer with the least possible surface area?* [Arranging the cubes in a solid 2 cm by 3 cm by 5 cm gives a total surface area of 62 cm². So does arranging them into a 3 cm by 3 cm by 3 cm cube and then placing the 3 other cubes on top.]

SOLUTIONS: DIVISION E

4E. For a number to leave a remainder of 5 when divided into 47, it must satisfy *two* conditions:
   1. it must divide exactly into 47 – 5, or 42, and
   2. it must be greater than 5, because the divisor is always greater than the remainder.

Strategy: List the numbers that satisfy one condition. See which also satisfy the other.
List all divisors of 42 in pairs. They are 1 and 42, 2 and 21, 3 and 14, 6 and 7. Of these, only 42, 21, 14, 6, and 7 are greater than 5. **There are 5 different counting numbers that will leave a remainder of 5 when divided into 47.**

Follow-Ups: *(1) What other number, besides 47, is less than 100 and leaves a remainder of 5 when divided by 6 or 7? [89] (2) What numbers between 100 and 300 leave a remainder of 5 when divided by 14 or 21? [131, 173, 215, 257, 299] (3) Why do the answers to #2 have the same interval, and why 42?*

5A. **METHOD 1:**   *Strategy: Use one cat's boots to replace the others' missing boots.*
The four boots for each cat had a total of 4 + 4 + 6 + 6 = 20 eyelets. The four missing boots are enough for one cat. The remaining boots are enough for the 3 other cats. **The total number of eyelets in the remaining boots is** 3 × 20 = **60.**

METHOD 2:   *Strategy: Consider the total number of eyelets for each cat.*
Each cat originally had four boots with a total of 20 eyelets. Two of the cats each now have a total of 20 – 4 = 16 eyelets in their boots and the other two cats, 20 – 6 = 14 eyelets each. The total number of eyelets in the remaining boots is 16 + 16 + 14 + 14 = 60.

5B. **METHOD 1:**   *Strategy: Examine the nature of multiples of 6 and 5.*
Two years ago my age was a multiple of 6 and therefore, even. The following year, (i.e. last year) my age was odd. An odd multiple of 5 must have a ones digit of 5. Then two years ago the ones digit of my age was 4. The only number less than 50 that has a ones digit of 4 and is a multiple of 6 is 24. Two years ago I was 24, last year I was 25, and **I am 26 now.**

METHOD 2:   *Strategy: Start with a table of multiples of 6.*
List the multiples of 6 that are less than 50 and determine which are 1 less than a multiple of 5.

My age 2 years ago	6	12	18	24	30	36	42	48
My age last year	7	13	19	25	31	37	43	49
Multiple of 5?	No	No	No	Yes	No	No	No	No

Last year I was 25, so I am 26 now.

Follow-Up: *Four years ago my age was a multiple of 5. Three years ago my age was a multiple of 4. Two years ago my age was a multiple of 3. How old am I? (Hint: Consider my age next year.) [59] Could the answer also be 119? [Very unlikely; it's an age.]*

SOLUTIONS: DIVISION E

5C. **METHOD 1:** *Strategy: For each choice made, consider the number of next choices.*
From Town A Kevin has a choice of 4 towns to visit. For each of these, he has a choice of 3 towns to go to next. The first two towns can be chosen in 4 × 3 or 12 ways. For each of these 12 choices, he has 2 choices for the third town to visit. There are 4 × 3 × 2 ways to visit the first 3 towns. For each of these 24 choices, he only has one choice for the last town to visit. **There are 4 × 3 × 2 × 1 = 24 different orders in which Kevin can visit the other four towns.**

METHOD 2: *Strategy: Make an organized list of all possible orders.*
The towns could be visited in the following orders:

Visit *B* first:	**BCDE**	**BCED**	**BDCE**	**BDEC**	**BECD**	**BEDC**
Visit *C* first:	**CBDE**	**CBED**	**CDBE**	**CDEB**	**CEBD**	**CEDB**
Visit *D* first:	**DBCE**	**DBEC**	**DCBE**	**DCEB**	**DEBC**	**DECB**
Visit *E* first:	**EBCD**	**EBDC**	**ECBD**	**ECDB**	**EDBC**	**EDCB**

There are 24 different orders in which Kevin can visit the other four towns.

METHOD 3: *Strategy: Make a partial list and generalize the result.*
As in Method 2, list the orders that begin with *B*. There are 6 of them. There will be the same number if Kevin starts with *C* or with *D* or with *E*. Then there are 6 × 4 = 24 different orders in which Kevin can visit the other four towns.

5D. **METHOD 1** *Strategy: Encase the figure in a BIG rectangle and subtract.*
Encasing a nonstandard figure in a rectangle offers a simple way to compute its area. The diagram shows the dimensions of the BIG rectangle as 23 cm by 20 cm, so its area is 460 sq cm. From this remove the four excess regions: 460 − 2 × (7 × 7) − 2 × (7 × 10) = 222 sq cm. **The total area of the figure is 222 sq cm.**

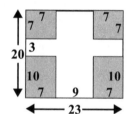

METHOD 2: *Strategy: Add the areas of the small rectangles.*

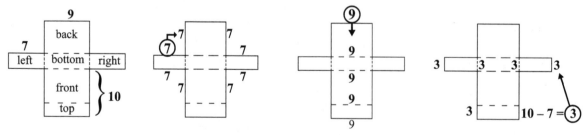

The area of both the back and front is 7 × 9 = 63 sq cm each. The area of both the bottom and top is 3 × 9 = 27 sq cm each. The area of both the left and right faces is 3 × 7 = 21 sq cm each. Adding the areas of the six rectangles, the total area of the figure is 222 sq cm.

SOLUTIONS: DIVISION E

5E. *Strategy: Use the test for divisibility by 9.*
Since 6*B*9 is divisible by 9, $6 + B + 9$ is divisible by 9. $B + 15$ is a multiple of 9. Therefore *B* is 3.

$$
\begin{array}{r}
5\,A\,4 \\
+\,1\,2\,5 \\
\hline
6\,3\,9
\end{array}
$$
The value of *A* is 1.

FOLLOW-UP: *Suppose that 6A7 and 9B2 are three-digit numbers, and that 6A7 is 275 less than 9B2. If 9B2 is divisible by 8, find all possible values for A.* [3 or 7]

SET 9 ◆ ◆ ◆ **Olympiad 1**

1A. *Strategy: Count each part of the line separately.*
There are 24 people in front of Kim and 11 people behind her. Counting Kim herself, **there are** $24 + 11 + 1 = $ **36 people in the line.**

1B. Denote the girls as *A*, *B*, *C*, *D*, and *E*.
METHOD 1: *Strategy: Consider the number of teams that include each girl.*
Designate the five girls as *A*, *B*, *C*, *D*, and *E*. Girl *A* can be teamed with any of the 4 other girls. Likewise, 4 teams include *B*, 4 teams include *C*, 4 teams include *D*, and 4 teams include *E*. However, this list of $5 \times 4 = 20$ teams lists each team twice. For example, *BC* is listed twice, as a team that includes *B* and as a team that includes *C*. Therefore there are $20 \div 2 = $ **10 different doubles teams** that **can be formed.**

METHOD 2: *Strategy: Make an organized list.*
All teams including *A*:   *AB, AC, AD, AE*
Teams including *B* but not *A*:   *BC, BD, BE*
Teams including *C* but not *A* or *B*: *CD, CE*
Teams not including *A*, *B*, or *C*:   *DE*
Counting up all the above, 10 different teams can be formed.

1C. **METHOD 1:** *Strategy: Use the symmetry of the given information.*
Notice that each of the costs mentioned is for a different pair of fruits, and each type of fruit is mentioned twice. Suppose Alex buys an apple and a pear, then a pear and a banana, and then an apple and a banana. His purchase of *two* of each kind of fruit would cost $25 + 19 + 16 = 60¢$. Then *one* of each would cost half as much, so **Alex spends 30¢.** It is worth observing that it is not necessary to find the cost of each type of fruit.

METHOD 2:   *Strategy: Determine the cost of one piece of each fruit.*

An apple and a pear cost 25¢, while a pear and a banana cost 19¢. Then one apple costs 6¢ more than one banana. So, for 6¢ more than the cost of one apple and one banana, Alex could buy *two* apples. Then 2 apples would cost 16 + 6 = 22¢, so one apple costs 11¢. Since one apple and one pear cost 25¢, one pear costs 25 − 11 = 14¢, and one banana costs 19 − 14 = 5¢. Therefore Alex spends 14 + 11 + 5 = 30 cents.

METHOD 3:   *Strategy: Set up a table. Guess and check.*

The first row of the table assigns different costs for one apple. The second and third rows use that number together with two of the given facts to compute the cost of the other fruits. The fourth row compares these amounts against the third given fact.

Suppose 1 apple costs:	8¢	9¢	10¢	11¢	12¢
Then 1 apple and 1 pear cost 25¢, so 1 pear costs:	17¢	16¢	15¢	14¢	13¢
Also 1 pear and 1 banana cost 19¢, so 1 banana costs:	2¢	3¢	4¢	5¢	6¢
From above, 1 apple and 1 banana cost:	10¢	12¢	14¢	16¢	18¢
Does 1 apple and 1 banana cost 16¢?	No	No	No	Yes	No

Alex spends 11 + 14 + 5 = 30¢.

FOLLOW-UP: A candy bar costs 25¢. If Amy, Beth, and Cara put their money together, they have 74¢. If Amy, Beth, and Dee put their money together, they have 72¢. Amy, Cara, and Dee have 71¢. Beth, Cara, and Dee have 68¢. If all 4 put their money together, how many candy bars can they buy? [3; 20¢ will be left over.]

1D. **METHOD 1:**   *Strategy: Change the figure to a simpler shape.*

Reduce the length (top and bottom) by 25 cm, leaving the shaded square. The perimeter of the square is 50 cm less than the perimeter of the original rectangle, which was 90 cm. The perimeter of the square is 40 cm and its side-length is 10 cm. The width of the original rectangle is 10 cm and the length is 10 + 25 = 35 cm. **The area of the rectangle is** 10 × 35 = **350 sq cm.**

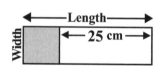

Alternatively, the width could be increased by 25 cm to form a square.

METHOD 2:   *Strategy: Find the semiperimeter.*

The perimeter of a rectangle, the sum of two lengths and two widths, is 90 cm. The sum of one length and one width, called the *semiperimeter*, is half of 90, or 45 cm. The two numbers with a sum of 45 that differ by 25 are 10 and 35. The area of the rectangle is 10 × 35 = 350 sq cm.

1E. _Strategy: Group conveniently and add the ones and tens digits separately._
The first 25 odd counting numbers are 1, 3, 5, 7, 9, …, 49. Arrange as shown in the table. Add the ones and tens digits separately in each column and then combine the subtotals.

The first 25 counting numbers	1	11	21	31	41	TOTALS
	3	13	23	33	43	
	5	15	25	35	45	
	7	17	27	37	47	
	9	19	29	39	49	
Sum of the ones digits	25	25	25	25	25	**125**
Sum of the tens digits	0	5	10	15	20	**50**
				Sum of all the digits		**175**

The sum of all the digits is 175.

**Follow-Ups:** (1) What is the 10th odd counting number? [19] The 50th? [99] The 2003rd? [4005] (2) What is the sum of the first 5 odd counting numbers? [25] The first 6? 7? 8? 101? [36; 49; 64; 10201]

SET 9   **Olympiad 2**

2A. _Strategy: Determine how much "blank wall" shows._
The picture requires 3 feet across so 22 feet of the wall is blank. The picture hangs in the center, so the blank wall on each side of the picture requires 11 ft. **It is 11 ft from the left edge of the wall to the left edge of the picture.**

2B. _Strategy: Work backwards, using inverse (opposite) operations._

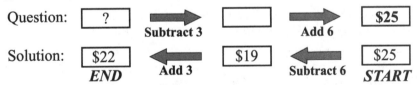

David paid $22 for the Beanie Baby.

2C. **METHOD 1:**   _Strategy: Guess Carlos' age. With a table find the other ages._

Choose values for Carlos' age. Use the facts that Ashley is twice as old as Carlos and Billy is 5 years younger than Ashley. For which value of Carlos' age is the sum of their ages 25? **Carlos is 6 years old.**

Trial Number	Carlos' age	Ashley's age	Billy's age	Sum	Sum = 25?
Trial 1	4	8	3	15	No
Trial 2	5	10	5	20	No
Trial 3	6	12	7	25	YES

METHOD 2:   *Strategy: Use algebra.*

Let Carlos' age = C. Then Ashley's age = 2C and Billy's age = 2C – 5.

(C) + (2C) + (2C – 5) = 25	First combine like terms.
5C – 5 = 25	Now add 5 to each side.
5C = 30	Divide by 5.
C = 6	Carlos is 6 years old.

2D.   METHOD 1:   *Strategy: Look for a pattern.*

Any two successive rows contain eight numbers. Each number is therefore 8 less than the number two rows above it. The entry 25 is 1 more than a multiple of 8. Every number that is 1 more than a multiple of 8 is in column T. Therefore **the number 25 belongs in column T.**

P	Q	R	S	T
	100	99	98	(97)
93	94	95	96	
	92	91	90	(89)
85	86	87	88	
	84	83	82	(81)
77	78	79	80	
				and so on …

METHOD 2:   *Strategy: Extend the table to list all numbers.*

P	Q	R	S	T
	100	99	98	97
93	94	95	96	
	92	91	90	89
⋮	⋮	⋮	⋮	⋮
	36	35	34	33
29	30	31	32	
	28	27	26	25

FOLLOW-UP: *Suppose Day 1 is a Tuesday. What day of the week is Day 2003? [Tuesday]*

2E.   METHOD 1:   *Strategy: Count the cubes that do **not** touch a side or the bottom of the box.*

Consider each of the 4 horizontal layers in turn. In the bottom layer of 16 cubes, each small cube touches the bottom of the box. In each of the other 3 layers, the 4 center cubes do not touch a side of the box. Then $3 \times 4 = 12$ cubes do not touch the bottom or a side of the box; therefore, $64 - 12 = $ **52 of the small cubes touch the bottom or a side of the box.**

Top View

METHOD 2:   *Strategy: Count the cubes that touch the box in each horizontal layer.*

In the bottom layer, all 16 cubes touch the box. In each of the three other layers, all the outer cubes touch either one or two sides of the box. That is 12 outer cubes per layer. Thus there are a total of $16 + 3 \times 12 = 52$ small cubes that touch the bottom or a side of the box.

FOLLOW-UPS: *(1) Suppose the box has a top. How many of the cubes do not touch the box? [8] (2) How many cubes do not touch a covered box which is completely filled with 1-cm cubes, if each side of the box measures 6 cm? 12 cm? n cm? [64; 1000; (n–2)³] (3) A covered box is filled with 5-cm cubes. All but one cube touches the box. How many cubes touch the box? [26]*

3A. **METHOD 1:** *Strategy: Find the total number of shirts.*
A total of $1 + 3 + 5 + 7 + 9 = 25$ shirts are bought, costing $8 each. **The five people spend $200 in total.**

METHOD 2: *Strategy: Find how much each person spends.*
At $8 per shirt, Amy spent $8 for one shirt, Becky spent $24 for 3 shirts, Colin spent $40 for 5 shirts, Dan spent $56 for 7 shirts and Emily spent $72 for 9 shirts. The five people spend $200.

(Note: The two methods are linked by the distributive property.)

FOLLOW-UPS: *An outfit consists of a shirt and a pair of pants. Shirts cost $8 each, and pants $24 each. What is the greatest number of outfits Amy can buy for $100 if: (1) each shirt can be worn with only one of the pairs of pants? [3] (2) any shirt can be worn with any pair of pants? [12]*

3B. *Strategy: Make a list and eliminate unwanted numbers.*
List all the counting numbers less than 25. Remove those divisible by 2 or by 5. Only the odd numbers, except for those ending in 5, remain: 1, 3, 7, 9, 11, 13, 17, 19, 21, 23. Their sum is 124. **The sum of all counting numbers less than 25 which are not divisible by 2 or 5 is 124.**

(Note: Of the 24 numbers, 12 numbers are divisible by two, and 4 numbers are divisible by five. However, 2 of those numbers are divisible by both two and five, and were counted twice. Then $12 + 4 - 2 = 14$ of the 24 numbers are divisible by two or five, and the remaining 10 numbers are not. There should be, and are, 10 numbers on the list above.)

3C. **METHOD 1:** *Strategy: Use a Venn diagram.*
Of the 8 runners, 6 are also mathletes. Then 2 girls are runners only. Of the 13 mathletes, 6 are also runners. Then 7 girls are mathletes only. Thus, a total of $6 + 2 + 7 = 15$ girls participate in at least one of the two activities. Of the total number of 25 girls, **10 girls are not on either team.**

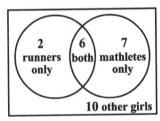

METHOD 2: *Strategy: Combine the teams but remove those counted twice.*
There are 13 mathletes and 8 runners. Of the $13 + 8 = 21$, 6 girls were counted twice. There is a total of $21 - 6 = 15$ girls on the teams, so $25 - 15 = 10$ girls are not on either team.

FOLLOW-UPS: *(1) Discuss the similarity between method 2 of 3C and the note following 3B. (2) Suppose 3 of the girls who are on both the math and basketball teams are also delegates to the student council, and only 5 of the girls in the original group of 25 are not on the student council or either team. How many girls are only on the student council? [5]*

SOLUTIONS: DIVISION E

3D. **METHOD 1:** *Strategy: Work backward, considering the numbers of steps.*

Third climb: After the second climb, $\frac{1}{8}$ of the steps remained. The number of steps remaining is therefore a multiple of 8. Assume the least numbers possible: 8 steps, of which I climbed 1.

Second climb: Here I climbed $\frac{1}{3}$ of the remaining steps, so the other $\frac{2}{3}$ was the 8 steps remaining after the second climb. Then the $\frac{1}{3}$ I climbed was 4 steps. Hence, $8 + 4 = 12$ steps remained after the first climb.

First climb: I climb half the total number of steps and 12 steps remain. Therefore, **the least possible number of steps in the staircase is 24.** This answer is correct because all numbers are whole numbers.

METHOD 2: *Strategy: Work forward, considering the fractions.*

First, I climb $\frac{1}{2}$ of the steps and $\frac{1}{2}$ of all the steps remain. Next, I climb $\frac{1}{3}$ of those remaining steps, which is $\frac{1}{3}$ of $\frac{1}{2}$ of all the steps; this is $\frac{1}{6}$ of all the steps. During the two climbs, I climbed $\frac{1}{2} + \frac{1}{6} = \frac{2}{3}$ of all the steps. Therefore, $\frac{1}{3}$ of all the steps remain after two climbs. Finally, I climb $\frac{1}{8}$ of this remaining $\frac{1}{3}$ of all the steps. That is $\frac{1}{24}$ of the total number of steps. The least number divisible by all the denominators is 24. Then the least number of steps in the staircase is 24. As a check, I climbed 12 steps at first, then 4 of the remaining 12 steps, and finally, 1 of the remaining 8 steps before stopping to catch my breath.

3E. **METHOD 1:** *Strategy: Find the area of the region that is not shaded.*

The area of rectangle $ABCD = 15 \times 8 = 120$ sq cm.
The area of rectangle $AEFD = 12 \times 8 = 96$ sq cm.
The area of unshaded triangle AEF = half of 96 = 48 sq cm.
The area of the shaded region = area of $ABCD$ – area of AEF.
The total area of the shaded regions is $120 - 48 = $ **72 sq cm.**

METHOD 2: *Strategy: Add the areas of the shaded regions.*

Rectangle $EBCF$: $EB = 15 - 12 = 3$ cm and $BC = 8$ cm. Area = 24 sq cm.

Triangle ADF: Area = $\frac{1}{2}$ of the area of rectangle $AEFD = \frac{1}{2}$ of $12 \times 8 = 48$ sq cm.

Shaded region: Area = the sum of the areas of rectangle $FCBE$ and triangle AFD.
The total area of the shaded regions is $24 + 48 = 72$ sq cm.

FOLLOW-UP: Find the area of the shaded region. [110 sq cm]

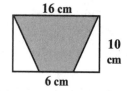

16 cm

10 cm

6 cm

SOLUTIONS: DIVISION E

4A. *Strategy: Draw a picture.*

The least number of pennies in one pile is 1. The next possibility is 2, and so on.
The sum of $1 + 2 + 3 + 4 + 5 + 6 = 21$. **The least possible total is 21.**

FOLLOW-UPS: There are 32 pennies in 6 piles. Each pile has a different number of pennies. (1) What is the greatest possible number of pennies in the tallest pile? [17] (2) What is the greatest possible number of pennies in the shortest pile? [2]

4B. *Strategy: Use knowledge of prime numbers.*
At least one of the ages must be even, because the sum of two odd numbers is an even number. The only even prime number is 2. Since 2 is also the least prime number, it must be Amanda's age. **Amanda is 2 years old.**

FOLLOW-UP: Suppose Carly is less than 50 years old. What is the greatest possible age she could be? [43]

4C. **METHOD 1:**   *Strategy: Use a ratio.*
24 adults require the same space as 30 children. Simpifying, 4 adults require the same space as 5 children. Since 5 more children could get on the bus to make 30 children, and 5 children require the same space as 4 adults, **4 adults can still get on the bus.**

METHOD 2:   *Strategy: Use fractions.*
The bus holds 30 children but 25 children are already on the bus. Then $\frac{25}{30} = \frac{5}{6}$ of the space on the bus holds children, leaving $\frac{1}{6}$ of the total space for adults. The bus holds 24 adults, so $\frac{1}{6}$ of them, 4 adults, can still get on the bus.

4D. *Strategy: Find the length of a side of the square.*
$AB + BC + CD + DE + EF + FG + GA = 180$.
$AB = BC = CD$ because they are sides of square $ABCD$.
$DE = GF$ because they are opposite sides of rectangle $DEFG$.
$DE = CD$, so $AB = BC = CD = DE = GF$.
$EF = DG$, so $AG + EF = AG + DG = AD$, the fourth side of square $ABCD$.
Then the sum of 6 equal lengths is 180 cm. Each of them must be 30 cm. $AG + 25 = 30$, so **the length of AG is 5 cm.**

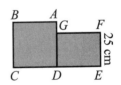

4E. **METHOD 1:** _Strategy: Find the total of each group of numbers._
In any set of numerical items, the average is their sum divided by the total number of items. Then the product of the number of items and the average is the sum. In the first set, the average of the 20 numbers is 20 and their sum is 400. In the second set, the average of the 60 numbers is 60 and their sum is 3,600. When the two sets are combined, the new set has 80 numbers with a sum of $400 + 3600 = 4000$. **The average of the combined group is** $4000 \div 80 = \mathbf{50.}$

(Note: Because the groups are of different size, the average of the combined groups is **not** the same as the average of the averages.)

METHOD 2: _Strategy: Use the ratio of the sizes of the two sets of numbers._
The combined group has 80 numbers. $\frac{1}{4}$ of them (20 out of 80) are in the first group and $\frac{3}{4}$ of them (60 out of 80) are in the second group. Then the average of the combined group (called a _weighted average_) is nearer to the average of 60 than it is to the average of 20, and $\frac{1}{4}$ of the interval from 60 to 20. $\frac{1}{4}$ of $60 - 20 = \frac{1}{4}$ of $40 = 10$. Thus the weighted average is 10 less than 60, which is 50.

(Note: Method 2 may be too sophisticated for some elementary students.)

FOLLOW-UP: _In the third marking period, the average of Kelsie's 7 math tests is 89. What grade does she need on the 8th test to raise her average to exactly 90?_ [97]

SET 9   **Olympiad 5**

5A. **METHOD 1:** _Strategy: Count backward from the first Sunday._
One week equals seven days. Earlier Sundays are May 18, 11, and 4. Therefore **May 1 occurs** three days earlier than May 4, **on Thursday.**

METHOD 2: _Strategy: Count forward from May 1._
May 1 is on the same day of the week as May 8, 15, and 22. May 22 is 3 days before May 25, and 3 days before Sunday is Thursday. May 1 occurs on a Thursday.

FOLLOW-UPS: _(1) Suppose there were only 5 days in a week (SMTWTh) and the 25th of the month fell on a Sunday. On what day would the 1st of the month be?_ [M] _(2) How many days could a week have for the 1st and the 25th to occur on the same day?_ [any factor of 24]

5B. _Strategy: Use the test for divisibility by 9._
If a number is divisible by 9, the sum of its digits is divisible by 9.
Then $A + 7 + A + 8$ must be 9 or 18 or 27 or 36.
$A + A + 15$ cannot be 9, because 9 is less than 15.
$A + A + 15$ cannot be 18, because A would not be a digit.
$A + A + 15$ cannot be 36, because A would not be a digit.
Thus $A + A + 15 = 27$ and **A represents 6.**

5C. <u>Strategy</u>: *Start with all fives and replace with twos and threes.*
In the table, a 25-point total is first achieved with five 5s. Then, in each case, the replacement has the same total point-score as the scores being replaced. The replacements are highlighted.

Start with:	five 5s, three 0s	5	5	5	5	5	0	0	0

Remove:	Replace with:	Result:							
one 5, one 0	one 2, one 3	5	5	5	5	2	0	0	3
two 5s, two 0s	two 2s, two 3s	5	5	5	2	2	0	3	3
three 5s, three 0s	three 2s, three 3s	5	5	2	2	2	3	3	3
two 5s, three 0s	five 2s	5	5	5	2	2	2	2	2
three 5s, two 0s	five 3s	5	5	3	3	3	3	3	0
four 5s, three 0s	one 2, six 3s	5	2	3	3	3	3	3	3

To replace five 5s requires more than 8 scores since 8 threes is only 24.
There are 7 combinations that give 25 points in 8 plays.

5D. <u>Strategy</u>: *Combine pieces to form more familiar shapes.*
Two of the "Els" can be placed next to each other to form a 2 × 4 × 2 rectangular solid.

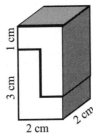

Three of these solids can be packed into a 2 × 4 × 6 box so that they fill it completely.

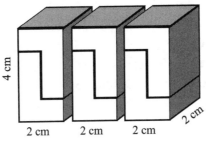

Therefore **6 "Els" can be packed in the box.**

5E. _Strategy: List the numbers that satisfy each condition. Look for a match._
Group first by sevens, since 7 is larger than 3. The numbers greater than five with 4 left over are: 11, 18, **25**, ... Beginning with 11, test each number on the list for a remainder of 1 upon division by 3. The least such number is 25. **The least number of marbles Kayla could have is 25.**

(Note: An alternate way is to make two lists, one for each condition, and look for a match.)

FOLLOW-UPS: _(1) What is the next three numbers of marbles that Kayla could have?_ [46; 67; 88] _(2) Do you notice a pattern?_ [Each number is 21 more than the last.] _Can you explain why the pattern holds?_ [To keep both remainders the same while increasing the quotients, the amount of increase must be a multiple of both numbers, i.e. a multiple of their LCM.] _(3) What is the 50th such number?_ [1033]

SET 10 ◆ ◆ ◆ **Olympiad 1**

1A. **METHOD 1:** _Strategy: Subtract in a useful way._
Note that each number in the first set is greater than its counterpart in the second set by an obvious amount. Thus:
$185 - 85 = 100$, $278 - 178 = 100$, and $579 - 279 = 300$.
Then $100 + 100 + 300 = 500$. **The value is 500.**

METHOD 2: _Strategy: Apply the order of operations._
Under the order of operations we evaluate the expression in parentheses first.
$(185 + 278 + 579) = 1042$.
$(85 + 178 + 279) = 542$.
$1042 - 542 = 500$. The value is 500.

1B. _Strategy: Determine what sides are opposite each other._
When folded, V is opposite X and U is opposite Y. The remaining face, labeled W, has no face opposite it. **W is on the bottom of the box.** Cutting out a model and folding it may make the solution easier to visualize.

1C. _Strategy: First decide who does not play the flute._
Alexis is a 6th grader. Emma and the flute player are 5th graders. Then Alexis does not play the flute, nor does Emma. Then Li plays the flute.

Alexis and the saxophone player practice together. Then Alexis does not play the saxophone. So Emma plays the saxophone and therefore **Alexis plays the drums.**

SOLUTIONS: DIVISION E

1D. **METHOD 1:** *Strategy: Form one, then two, then three triangles ...*
The first three toothpicks form one triangle. After that, every two additional toothpicks form another triangle. After the first three toothpicks are used, 86 are left. These 43 pairs of toothpicks form 43 more triangles. Including the first triangle, **44 triangles are formed**.

METHOD 2: *Strategy: Make a table of simpler cases and look for a pattern.*

Number of toothpicks	3	5	7	9	11	...	89
Number of triangles	1	2	3	4	5	...	?

If we subtract 1 from the number of toothpicks and then divide by 2, the result is the number of triangles. $89 - 1 = 88$ and $88 \div 2 = 44$, so 89 toothpicks form 44 triangles. (Some students may obtain 44 triangles by finding other patterns in this table.)

FOLLOW-UPS: (1) Suppose the toothpicks are used to form a row of 100 squares, one toothpick per side. How many toothpicks are needed? [301] (2) How many are needed to form a row of 100 pentagons in a similar manner? [401] (3) You may want to extend this pattern through polygons of 6, 8, 10 sides, etc.

1E. *Strategy: Find the time needed for each part of the trip.*
The only way to find Dale's average rate for the whole trip is to divide the <u>total</u> distance traveled by the <u>total</u> time needed.

From A to B is 120 miles. Traveling at 60 mph, Dale takes $120 \div 60 = 2$ hours. Similarly, in traveling from B to C, Dale takes $120 \div 40 = 3$ hours, and the trip from C back to A takes $120 \div 24 = 5$ hours. The total distance is $3 \times 120 = 360$ miles. The total time is $2 + 3 + 5$ or 10 hours. **Dale's average rate for the entire trip is** $360 \div 10 = $ **36 mph.**

(Note: Why is the answer unequal to the average of 60, 40, and 24 mph? In general, if the times are different, then the average rate for the entire trip is different from the average of the individual rates. For example, consider an extreme case: 50 miles traveled at 2 mph followed by 50 miles traveled at 50 mph. Because the first trip alone requires 25 hours, the average rate for the entire 100 miles cannot be 26 mph.)

FOLLOW-UPS: The numbers of miles between cities A, B, and C are shown in the diagram. Dolly bikes from A to B at 25 mph and from B to C at 10 mph. (1) How fast does she travel from C to A if her average rate for the entire trip is 16 mph? [20 mph] (2) Dolly wants her average rate for the entire trip to be 20 mph. Can she do this? Why? [No. She would have to complete the trip in a total of 6 hours, and she has already used that much time.]

2A. **METHOD 1:** *Strategy: Find the least common multiple.*
Any number divisible by both 3 and 5 is a multiple of their LCM, 15. An even multiple of 15 is a multiple of 30. The only multiple of 30 between 100 and 125 is 120. **The number is 120.**

METHOD 2: *Strategy: Consider one condition at a time.*
The numbers between 100 and 125 that are divisible by 5 are 105, 110, 115, and 120. Of these four, only 110 and 120 are even. Of these two, only 120 is divisible by 3. The number is 120.

FOLLOW-UP: How many odd counting numbers between 500 and 700 are divisible by 3 and 5? [6]

2B. **METHOD 1:** *Strategy: Determine the weekly decrease in the difference.*
Each week Dom adds $6 and Kim withdraws $4, so their account totals become closer by $10. Kim starts with 200 – 120 = $80 more than Dom, so **at the end of** $80 \div 10 =$ **8 weeks they will have the same amount in their accounts.**

METHOD 2: *Strategy: Build a table of the weekly amounts in each account.*

Week #	Start	Amount at the end of week							
		1	2	3	4	5	6	7	8
Kim	200	196	192	188	184	180	176	172	168
Dom	120	126	132	138	144	150	156	162	168
Differences	80	70	60	50	40	30	20	10	0

They have the same amount at the end of 8 weeks.

METHOD 3: *Strategy: Use algebra.*
Let w = the number of weeks that pass.
Dom has $120 + 6w$ dollars and Kim has $200 - 4w$ dollars at the end of w weeks.
$$120 + 6w = 200 - 4w$$
$$10w = 80$$
$$w = 8$$
At the end of 8 weeks they each have the same amount.

2C. *Strategy: Find the total number of points earned.*
In each race $5 + 3 + 1 = 9$ points are earned. The three racers earn a total of $3 \times 9 = 27$ points. If each racer has the same score, each would have 9 points. Thus, Beth's total score is greater than 9. Each runner's total is the sum of three odd numbers and therefore is odd. The least odd number greater than 9 is 11, which is a possible winning score. For example, the scores could be 5, 5, 1; 3, 3, 3; 1, 1, 5. **Beth's least possible total score is 11.**

FOLLOW-UP: (1) 6 darts land on a dartboard. Each scores 5 or 7 or 9 points. Which score of 24, 33, 39, 46, 51, and 60 is possible? [46] (2) How many different ways can the total score of 46 be reached? [3 (777799, 577999, 559999: exchange two 7s for one 5 and one 9]

2D. _Strategy: Examine the totals of the three rows._

From top to bottom the totals of the three rows are 46, 39, and 53. The total for the second row is 7 less than that for the first row, and the total for the third row is 7 more than that for the first row. To make the three totals equal, increase the total of the second row by 7 and decrease the total of the third row by 7. The only two numbers in these rows that differ by 7 are 4 and 11. Once **4 and 11 are exchanged**, the total of each row becomes 46.

**Follow-Up:** A magic square is an arrangement of numbers so that the total of each row and each column, as well as the total of the two main diagonals, is the same. Complete the following magic square inserting each odd number from 3 through 19. [From left to right the rows contain 17,3,13; 7,11,15; and 9,19,5.]

	3	
		5

2E. _Strategy: Trace possible paths on the cube._

Each possible path has a length of 4, 6, or 8 m. **The longest such path is 8 m.** One such path is shown below.

SET 10   **Olympiad 3**

3A. _Strategy: Draw a diagram._

Draw a line and place the points on it using the information given.

Statement	Diagram
1. _A_ is between _B_ and _C_.	•—•—• or •—•—•
	_B A C_ _C A B_
2. _B_ is between _A_ and _D_.	•—•—•—• or •—•—•—•
	_D B A C_ _C A B D_
3. _D_ is to the left of _C_.	•—•—•—•
	_D B A C_

The points in order from left to right are _DBAC_.

3B. METHOD 1:   _Strategy: Start with an extreme case._

Suppose Jeffrey did additional chores on _none_ of the ten days. He would have earned $30. This is $6 less than he actually earned. Each day he does his additional chores he earns an additional $2, so **he did his additional chores on 3 days.**

METHOD 2:   _Strategy: Build a table._

The first line of the table shows the number of days Jeffrey does only those certain chores and the second line shows the remainder of the ten days, in which he does additional chores. The third line shows the total amount of money he earns.

NUMBER OF DAYS

The certain chores	0	1	2	3	...	?
Additional chores	10	9	8	7	...	?
Total dollars earned	50	48	46	44	...	36

Using the pattern of the third line, we can find that he did additional chores on three days.

FOLLOW-UP: Farmer Jones has chickens and goats. When he counts these animals' heads, the total is 40. When he counts these animals' feet, the total is 100. How many chickens does he have? [30]

3C. _Strategy: Apply one divisibility rule at a time._

Rule	**Possible Numbers**
1. A number divisible by 5 ends in 5 or 0.	8 6 _**0**_ _or_ 8 6 _**5**_
2. Only even numbers are divisible by 4.	8 6 _**0**_
3. If a number is divisible by 3, the sum of its digits is divisible by 3.	8 6 **1** 0 _or_ 8 6 **4** 0 _or_ 8 6 **7** 0
4. If a number is divisible by 4, the number formed from its last 2 digits is divisible by 4.	86**40**

The four-digit number is 8640.

FOLLOW-UP: 8640 is also divisible by 12 and 15. Explain how you can know this without dividing. [Any number divisible by 3 and 4 is also divisible by 12. Any number divisible by 3 and 5 is also divisible by 15.]

3D. _Strategy: Split the region into simpler figures._

Split the region _ABCD_ into rectangle _ABFD_ and triangle _BFC_ as shown. Complete rectangle _BFCE_. The area of _ABFD_ is 6 × 4 = 24 sq cm. The area of triangle _BFC_ is half the area of rectangle _BFCE_. Because _BF_ = _AD_, _BF_ is 6 cm. To find _FC_, observe that _DF_ = _AB_ = 4 cm and _FC_ = _DC_ − _DF_. Then _FC_ = 8 − 4 = 4 cm. Since the area of _BFCE_ = 6 × 4 = 24 sq cm, the area of triangle _BFC_ = $\frac{1}{2}$ × 24 = 12 sq cm. **The area of _ABCD_ is 24 + 12 = 36 sq cm.**

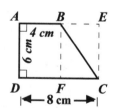

3E. _Strategy: Determine the difference gained each hour._

Every 2 hours one watch gains 3 minutes and the other watch loses 1 minute. Then for every 2 hours the difference between their times grows by 4 minutes. For every hour the difference gained is 2 minutes. The faster watch will gain 60 minutes on the slower one in $60 \div 2 = 30$ hours. Thirty hours after 7 AM is 1:00 PM the next day. **The faster watch will be one hour ahead of the slower watch at 1:00 PM the next day.**

**Follow-Ups:** (1) Suppose another fast watch is set correctly at 7 PM and gains 4 minutes every hour. To the nearest minute, what is the correct time when the watch first shows 3:00? [2:30 AM] (2) Suppose a watch gains 2 minutes per hour. If the watch reads correctly on Tuesday, January 13 at 7 PM, what is the next day and time in which the watch reads correctly? [Thursday, January 28 at 7 PM]

SET 10   **Olympiad 4**

4A. **METHOD 1:**   _Strategy: Use the test for divisibility by 25._

Any whole number ending in 25, 50, 75, or 00 is a multiple of 25. The least such number greater than 259 is 275, which is 16 more than 259. **The least counting number that can be added to 259 is 16.**

METHOD 2:   _Strategy: List the multiples of 25 greater than 200._

If 259 is divided by 25, the remainder is 9. Add 16 to 9 to reach the next multiple of 25, which is $259 + 16 = 275$. The least counting number is 16.

**Follow-Up:** Suppose $259 - N$ is a multiple of 40. What is the least possible value of N? [19]

4B. _Strategy: Count the favorable outcomes._

The factors of 6 are 1, 2, 3, and 6. These are the four favorable outcomes. The top face could show 1, 2, 3, 4, 5, or 6, a total of six possible outcomes. All outcomes are equally likely, so the probability of a favorable outcome is the ratio of the number of _favorable_ outcomes to the number of _possible_ outcomes. **The probability that the top face shows a factor of 6 is $\frac{4}{6}$ or $\frac{2}{3}$. Other acceptable answers are: 4 out of 6; 2 out of 3; or $66\frac{2}{3}$ %.**

**Follow-Up:** A fair die is thrown. What is the probability that: (1) the top face shows a prime number? $[\frac{1}{2}]$ (2) the sum of the five visible faces (top and four sides) is a prime number? $[\frac{1}{3}]$

SOLUTIONS: DIVISION E

4C. **METHOD 1:**   *Strategy: Reason from right to left.*

1. From the given choices, only $D = 8$ and $H = 9$ can produce 9. Regrouping occurs.

2. Because of regrouping, C and G must be the same digit. From the choices, $C = G = 3$. Regrouping occurs.

3. Because of regrouping, either B or $B + 10$ is 7 more than F. From the remaining choices, only $B = 4$ and $F = 7$ can produce 6. Regrouping occurs.

4. By elimination, $A = 5$ and $E = 1$. Thus $ABCD = 5438$ and $EFGH = 1739$. **The squares represent the four-digit number 5438.**

```
1.  A B C 8
   -E F G 9
    3 6 9 9

2.  A B 3 8
   -E F 3 9
    3 6 9 9

3.  A 4 3 8
   -E 7 3 9
    3 6 9 9

4.  5 4 3 8
   -1 7 3 9
    3 6 9 9
```

METHOD 2:   *Strategy: Change subtraction to addition.*

It is clearer and easier to follow the reasoning if we add rather than subtract. However, the reasoning is very similar in both cases.

```
   E F G H
  +3 6 9 9
   A B C D
```

4D. *Strategy: Split the region into simpler figures.*

Split the region into nine congruent squares, as shown. Each square has an area of $3 \times 3 = 9$ cm^2.

METHOD 1:   *Strategy: Find the area that is not shaded.*

The area of the large square is $9 \times 9 = 81$ cm^2. Each of the four unshaded triangles is half of one of the small squares. The four triangles together have an area equal to two of the small squares. The unshaded area is then $2 \times 9 = 18$ cm^2, and **the area of the shaded region is** $81 - 18 = $ **63 sq cm.**

METHOD 2:   *Strategy: Add the areas of the smaller shaded regions.*

The four shaded triangles are equal in area to two of the small squares. Combined with the five small shaded squares, the shaded region is equal in area to a total of seven small squares. The total area shaded is then $7 \times 9 = 63$ sq cm.

4E. *Strategy: Count in an organized way.*

Listing of beads	Total number of red beads	Total number of all beads
RG	1	2
RG RRGG	$1 + 2 = 3$	6
RG RRGG RRRGGG	$1 + 2 + 3 = 6$	12
⋮	⋮	⋮
RG ... RRRRRRRRRGGGGGGGGG	$1 + 2 + 3 + ... + 9 = 45$	90

The first nine groups contain 45 beads of each color. The tenth group starts with 10 red beads, and that brings the total number of beads to 100. **Of the first 100 beads, 55 are red.**

FOLLOW-UP: How many green beads are there in the first 200? [95]

5A. **METHOD 1:**   *Strategy: Find the prime factors and recombine them.*
$6 = 2 \times 3$ and $150 = 5 \times 30$ which can be written as $5 \times 5 \times 2 \times 3$.
Then $6 \times 150 = (2 \times 3) \times (5 \times 5 \times 2 \times 3)$. Observe that each prime factor occurs twice.
Rearrange the factors: $(2 \times 3 \times 5) \times (2 \times 3 \times 5)$, which becomes 30×30.
30 multiplied by itself is equal to the product of 6 and 150.

METHOD 2:   *Strategy: Do the multiplication.*
$6 \times 150 = 900$. Since $9 = 3 \times 3$, then $900 = 30 \times 30$. The number is 30.

FOLLOW-UP: A large cube is built using all the small cubes from 25 packages. Each package contains 40 one-cm cubes. How high is the large cube? [10 cm]

5B. **METHOD 1:**   *Strategy: Combine one stamp of each kind into a "super stamp".*
Imagine one "super stamp" consisting of one of each of the stamps mentioned. This super stamp costs $50 + 20 + 10 + 5 = 85¢$. There are $510 \div 85 = 6$ of these super stamps and therefore **Jessie has 6 fifty-cent stamps.** This method illustrates the distributive property.

METHOD 2:   *Strategy: Make a table, starting with simpler cases.*
The table shows the computation of values for 1, 2, 3, ... sets of stamps.

Number of each kind of stamp	1	2	3	4	5	6
Value of 50¢ stamps, in cents	50	100	150	200	250	300
Value of 20¢ stamps, in cents	20	40	60	80	100	120
Value of 10¢ stamps, in cents	10	20	30	40	50	60
Value of 5¢ stamps, in cents	5	10	15	20	25	30
Total value, in dollars	$0.85	$1.70	$2.55	$3.40	$4.25	**$5.10**

Jessie has 6 fifty-cent stamps. Observe that the row of the table showing the total value verifies that super stamps cost 85¢ each.

FOLLOW-UP: Jessie has stamps worth 7¢, 13¢, or 37¢ each. She has the same number of two kinds of these stamps and a greater number of the third kind. The total value of the stamps is $2.49. How many of each kind does she have? [Two solutions, each requiring 15 stamps: (7@7¢, 4@13¢, 4@37¢); (3@7¢, 9@13¢, 3@37¢)]

5C. _Strategy: Simplify the figure._

Think of the given figure as a rectangle with two smaller rectangles cut out of it as shown in *figure 1*. Since opposite sides of a rectangle are equal in length, the given figure will have the same perimeter as the rectangle shown in *figure 2*. Since the perimeter of the rectangle is $2 \times (10 \text{ m} + 11 \text{ m})$, **the perimeter of the given figure is 42 meters.**

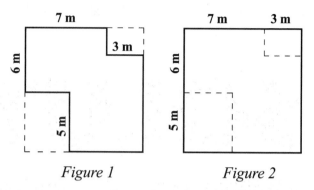

Figure 1   Figure 2

(Note: Some people visualize figures 1 and 2 as walls of a building before and after a renovation, respectively. They say some walls were simply pushed out.)

5D. _Strategy: Find the possible sums of the numbers in the horizontal row._

Note first that the sum of all the numbers $1 + 2 + \ldots + 9$ is 45. If the sum of the *five* numbers in the row equals the sum of the *five* numbers in the column, then the sum of the *four* numbers in each, excluding M, must also be equal. The sum of the four numbers in the row must be less than 22.5, which is half of 45. $4 + 9 + 7 = 20$, so the sum of the four numbers in the row must be 21 or 22. This is also the sum of the four numbers in the column. All eight squares, except M, have a sum of either 2×21 or 2×22. Then M is either $45 - 42$ or $45 - 44$. **M can be 1 or 3, only.**

(If $M = 1$, both sums are 23, the numbers in the row are 1, 4, 9, 7, 2, and the numbers in the column are 1, 3, 5, 6, 8. If $M = 3$, both sums are 24, the numbers in the row are 3, 4, 9, 7, 1 and the numbers in the column are 3, 2, 5, 6, 8.)

FOLLOW-UP: *Place 2, 3, 4, 5, and 7 in the five boxes so that the sum of B, C, and D equals the sum of A, C, and E. (1) Which values are possible for C? [3,7 only] (2) How many different ways in all can the letters be replaced by these five numbers so that the two sums are equal? [16]*

5E. **METHOD 1:** _Strategy: Find the total number of marbles in the bag._

The probability of drawing a red marble is $\frac{2}{3}$, so the probability of drawing a black marble is $1 - \frac{2}{3} = \frac{1}{3}$. Then 8 is $\frac{1}{3}$ of the total number of marbles. There are $3 \times 8 = 24$ marbles in all. Since 8 of them are black, **there are 16 red marbles in the bag.**

METHOD 2: _Strategy: Find the ratio of the numbers of black and red marbles._

The probabilities are $\frac{2}{3}$ for a red marble and $\frac{1}{3}$ for a black marble. Thus, there are twice as many red marbles as black ones. Since the bag has 8 black marbles, it must have 16 red marbles.

FOLLOW-UP: *Suppose the bag contains red, white, and blue marbles and no other color. The probability of drawing a white marble is $\frac{1}{10}$. The probability of drawing a red marble is twice that of drawing a white marble. The bag contains 14 blue marbles. How many red marbles are in the bag? [4]*

SOLUTIONS, STRATEGIES, & FOLLOW-UPS

DIVISION M
SETS 11-17

SOLUTIONS: DIVISION M

1A. METHOD 1: _Strategy: Consider pairs of opposite faces._

Fold the figure mentally into a cube. Pairs of opposite faces: 1 and 3, 2 and 6, 4 and 5. The sum of the greater number of each pair yields $3 + 6 + 5 = 14$. **The greatest sum of three numbers whose faces meet at a corner is 14.**

METHOD 2: _Strategy: Search for the greatest sum without regard to corners._

Fold the figure mentally into a cube. The greatest possible sum is $6 + 5 + 4 = 15$. But while the faces containing 6 and 5 have a corner in common, the faces containing 4 and 5 cannot touch because they are opposite each other. The next greatest possible sum is $6 + 5 + 3 = 14$. The sum of 14 works because 3 shares a corner with both 5 and 6.

1B. METHOD 1: _Strategy: Average the least and greatest multiples._

List the set of multiples in order. It is an arithmetic sequence (one with a constant difference between terms). The average of the first and last terms is $(3 + 99) \div 2 = 51$. The average of the second and next-to-last terms is $(6 + 96) \div 2 = 51$. The average of the third and second-to-last terms is $(9 + 93) \div 2 = 51$. In fact, the average of every pair of corresponding terms is 51. Therefore, **the average of the entire set of multiples is 51.** The easiest way to find the average of such a set of numbers is just to average the first and last terms.

METHOD 2: _Strategy: Use the definition of average._

By definition, the average is the sum of all the multiples divided by the number of multiples. The number of multiples is 33, because 3 is the first multiple and 99 is the 33rd. The sum is 1683, and the average is $1683 \div 33 = 51$.

An easier way to find the sum is to use an approach often attributed to Gauss, by writing the sequence horizontally twice and adding vertically.

$$\begin{aligned} \text{sum} &= 3 + 6 + 9 + \ldots + 96 + 99 \quad \text{and again as} \\ \text{sum} &= 99 + 96 + 93 + \ldots + 6 + 3 \quad \text{adding yields} \\ 2 \times \text{sum} &= 102 + 102 + 102 + \ldots + 102 + 102 \quad \text{(102 appears 33 times)} \end{aligned}$$

Each 102 is the sum of two multiples. Since each number was counted twice in the addition, the sum of all the multiples is half of 33×102, or just 33×51. Then the average of all the multiples is $(33 \times 51) \div 33 = 51$. Observe that in this method, very little computation was done.

FOLLOW-UP: _It is thought that the great mathematician Karl Friedrich Gauss, as a nine-year-old, was asked by his teacher to add all the counting numbers from 1 to 100 inclusive. The tedious task was intended to keep him occupied for several minutes, but he startled the teacher by answering within seconds. Method 2 is one approach he may have used. Can you find another way? Research may help._

1C. **METHOD 1:** *Strategy: Examine the differences between scores.*
The least sum is $3 \times 2 = 6$ and the greatest sum is $3 \times 8 = 24$. The 5-ring scores 3 points more than the 2-ring and the 8-circle scores 3 points more than the 5-ring. Thus "moving" a dart from the 2-ring increases the sum by a multiple of 3. Hence, all possible sums are multiples of 3 and are between 6 and 24, inclusive; **7 sums are possible.**

METHOD 2: *Strategy: List the possibilities in an organized way.*
Here, the solver must be careful to list each possible outcome exactly once, so a systematic method of enumeration is required. An organized list works well. Consider scores in which:

A. three scores are equal

$8 + 8 + 8 = \boxed{24}$
$5 + 5 + 5 = \boxed{15}$
$2 + 2 + 2 = \boxed{6}$

B. two scores are equal

$8 + 8 + 5 = \boxed{21}$
$8 + 8 + 2 = \boxed{18}$
$5 + 5 + 8 = 18$
$5 + 5 + 2 = \boxed{12}$
$2 + 2 + 8 = 12$
$2 + 2 + 5 = \boxed{9}$

C. No scores equal

$8 + 5 + 2 = 15$

There are 7 different possible sums.

FOLLOW-UP: How many different sums are possible if 4 darts land? 5? 6? 100? [9; 11; 13; 201]

1D. *Strategy: Use the definition of a perfect cube.*
A perfect cube is the product of three identical factors. For example, $8 = 2^3 = 2 \times 2 \times 2$. Consider the prime factors of $45 = 3^2 \times 5$. Since $45N$ must be the smallest perfect cube, then N must contribute a *minimum* number of additional factors to the product to make $45N$ a cube. One additional factor of 3, and two additional factors of 5 will produce the least cube. Then $N = 3 \times 5 \times 5 = 75$. **The least whole number value of N is 75.**

FOLLOW-UPS: What is the next whole number value of N? [600] *Of what number is it the cube?* [30]

1E. *Strategy: Rotate the lower square about the center of the upper square.*
The diagonals meet at the center of the upper square. *Figure 1* shows the lower square rotated about this center. The area of the shaded region is one-fourth of the area of the upper square. Thus, since the area of the upper square is 100 sq cm, the area of the shaded region in *figure 1* is 25 sq cm.

Figure 1

To see that the rotated shaded region is equal in area to the original shaded region, use *figure 2*. If we draw perpendicular line segments from the center to the sides of the upper square, we form two identical triangles, equal in area. Thus, both the original and the rotated shaded regions are equal in area and **the area of the original shaded region is 25 sq cm.**

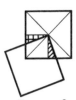

Figure 2

2A. **METHOD 1:** *Strategy: Use the fact that opposite sides of a rectangle are equal.*
Because all seven rectangles are congruent, $AF = FB$. Thus, $\overline{BA}$ is divided into two congruent parts and the length of each is 10. Therefore, $ED = 10$ and $GC = 10$, also.

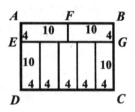

Since $AB = 20$, $DC = 20$. Similarly, $\overline{DC}$ is divided into five equal parts, the corresponding sides of five congruent rectangles. The length of each part is 4. Thus, each of the congruent rectangles are 10 by 4. Hence, $AE = 4$ and $BG = 4$.

Combining AE and ED, $AD = 14$, and therefore, $BC = 14$. **The perimeter of rectangle ABCD is** $20 + 14 + 20 + 14 = \mathbf{68}.$

METHOD 2: *Strategy: Use algebra.*
Since each of the seven rectangles are congruent, they have equal widths. Represent the width of each smaller rectangle by x, and the length of each smaller rectangle by y. Then $AB = 20 = 2y$, so $y = 10$. Since CD is the sum of 5 equal widths, $CD = x + x + x + x + x = 5x$. Then $CD = 20 = 5x$, so $x = 4$. Then $AD = x + y = 14$ and $AB = 20$, so the perimeter of rectangle $ABCD$ is $2(20 + 14) = 68$.

FOLLOW-UP: Suppose the perimeter of rectangle ABCD is 102. What is its area? [630]

2B. *Strategy: Find the length of each cycle, then compute the Least Common Multiple.*
The word "MATH" appears every **fourth** row, beginning with row 1: 1, 5, 9, 13, 17, …

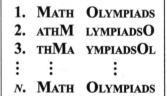

"OLYMPIADS" appears every **ninth** row, also beginning with row 1: 1, 10, 19, 28, …

Using the Least Common Multiple LCM (4,9) = 36, it can be seen that both "MATH" and "OLYMPIADS" appear every 36 rows. In the first row (and *only* the first row) of each set of 36, both words are spelled correctly. The next correct spelling appears in the first row of the second set. The next correct spelling of "MATH OLYMPIADS" appears in row 37, so $n = \mathbf{37}$.

FOLLOW-UPS: (1) In this problem, find the value of n in the tenth occurrence. [325] *(2) Can you develop a formula to compute any occurrence? (3) Two-meter-long sections of fence cost $100 each and posts cost $20 each. What is the cost of erecting a 30-meter-long fence?* [$1820; there is one more post than section.]

2C. **METHOD 1:**   *Strategy: Apply the test of divisibility for 5.*

Examine $109 - 5P = 7Q$. The product $5P$ is a multiple of 5 and ends in 0 or 5. If $5P$ is now subtracted from 109, the difference ends in the digit 4 or 9. Therefore $7Q$, a multiple of 7, is 14 or 49 or 84. Subtracting each of these from 109 produces $5P = 95$ or 60 or 25. Dividing by 5, we obtain $P = 19$, 12, or 5. Since both P and Q must be prime, it is necessary to compute the corresponding values of Q. If $P = 19$, then $7Q = 14$ and $Q = 2$. To be complete, the other two cases should be examined. Eliminate $P = 12$ since 12 is not prime. If $P = 2$, then $7Q = 84$ and $Q = 12$ which is not prime. Therefore, **the value of the prime P is 19.**

METHOD 2:   *Strategy: Examine all terms for parity (odd vs. even).*

Since 109 is odd, either $5P$ or $7Q$ must be even. Examine each possibility.

Case 1: Suppose $5P$ is even; then P is even. However the only even prime is 2, making $P = 2$. Then $5(2) + 7Q = 109$ and $7Q = 99$, which is impossible.

Case 2: Suppose $7Q$ is even; then $Q = 2$. Then $5P + 7(2) = 109$ and $5P = 95$. Dividing by 5 yields $P = 19$. Both $P = 19$ and $Q = 2$ are prime. P must equal 19.

2D. *Strategy: Use the division algorithm and examine the remainders.*

Fact: The division algorithm asserts that the remainder in a division of whole numbers is always less than the divisor. In this problem, the greatest remainder could be at most one less than the sum of the digits. To maximize the sum, we choose digits as great as possible. Test each case from the greatest sum of the digits down:

sum of digits = 18: $99 \div 18 = 5\ R\ 9$ $\Rightarrow$ remainder 17 is not obtained.

sum of digits = 17: $\left.\begin{array}{l} 98 \div 17 = 5\ R\ 13 \\ 89 \div 17 = 5\ R\ 4 \end{array}\right\}$ $\Rightarrow$ remainder 16 is not obtained.

sum of digits = 16: $\left.\begin{array}{l} 97 \div 16 = 6\ R\ 1 \\ 79 \div 16 = 4\ R\ \mathbf{15} \\ 88 \div 16 = 5\ R\ 8 \end{array}\right\}$ $\Rightarrow$ **The greatest remainder obtained is 15.**

2E. **METHOD 1:**   *Strategy: Draw a line segment to form two smaller rectangles.*

Drop a perpendicular from E to meet side BC at point F. Now we have three rectangles! $\overline{BE}$ and $\overline{EC}$ are two of their diagonals. A diagonal divides a rectangle into two congruent (identical) triangles. $\triangle AEB \cong \triangle FBE$ and $\triangle EDC \cong \triangle CFE$, so $\triangle BEC$ has exactly one-half the area of rectangle ABCD. Since the area of $\triangle BEC = \frac{1}{2} \times 6 \times 8 = 24$ sq units, **the area of rectangle $ABCD$ is 48 square units**.

SOLUTIONS: DIVISION M

METHOD 2: *Strategy: Represent the area of triangle BEC two ways and equate.*

Draw $\overline{EF} \perp \overline{BC}$. To find the area of $\triangle BEC$, either consider perpendicular segments $\overline{BE}$ and $\overline{EC}$ as the base and height, or consider $\overline{BC}$ and $\overline{EF}$ as the base and height. Then the area of the triangle equals either $\frac{1}{2} \times BE \times EC$ or $\frac{1}{2} \times BC \times EF$. Then $BE \times EC = BC \times EF$, since both multiplications produce the same area. Because $BE \times EC = 8 \times 6 = 48$, then $BC \times EF = 48$. But $\overline{BC}$ is the length of the rectangle and $\overline{EF}$ is equal to $\overline{DC}$, the width of the rectangle. Therefore, the area of the rectangle is 48 square units.

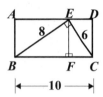

3A. METHOD 1: *Strategy: Examine the nature of each of the two sets of numbers.*

From fact (2), N is 3 more than a multiple of 5 and therefore ends in 3 or 8. From fact (1), N is 1 more than a multiple of 4, and is therefore odd. N ends in 3. List numbers greater than 40 that end in 3 and test them against both facts: 43, 53, 63, … **The least value of N is 53.**

METHOD 2: *Strategy: List both sets and compare.*

Fact (1) produces the set 41, 45, 49, 53, 57, 61, 65, 69, 73, 77, 81, ...
Fact (2) produces the set 43, 48, 53, 58, 63, 68, 73, ...
The first entry to appear in both sets is 53. The smallest value of N is 53.

FOLLOW-UP: List the next 3 whole numbers that satisfy the conditions of the problem.[73, 93, 113] *Why do they differ by 20?*

3B. METHOD 1: *Strategy: Eliminate the impossible numbers of 25¢ stamps.*

Any multiple of 16 is even and 362 is even, so the multiple of 25 is even. Then the 25¢ stamps cost Josephine an amount ending in 50¢ or 00¢. As a result, the 16¢ stamps cost her an amount ending in 12 or 62¢. The least such multiple of 16¢ is $1.12, the cost of 7 stamps. **Josephine purchased 7 of the 16¢ stamps.**

METHOD 2: *Strategy: Make an organized chart.*

Since $3.62 \div 25 \approx 14$, start with the maximum number of 25¢ stamps, 14, and count downwards. Organize results in a table as shown. Since $1.12 \div 16¢ = 7$, there are 7 of the 16¢ stamps.

Variation: In the first column list the numbers of 16¢ stamps and in the fourth column check whether or not the change is divisible by 25.

Number of 25¢ stamps	Total Cost	Change From $3.62	Divisible by 16?
14	3.50	.12	No
13	3.25	.37	No
12	3.00	.62	No
11	2.75	.87	No
10	2.50	1.12	YES

3C. <u>METHOD 1:</u>   *Strategy: Determine the number of "student-classes" required.*
There are 1600 students × 5 classes each = 8000 "student-classes" daily. There are 8000 student-classes ÷ 25 students per class = 320 classes daily. Finally, 320 classes ÷ 4 classes per teacher = **80 teachers in Euclid City School**.

<u>METHOD 2:</u>   *Strategy: Determine the number of students for each teacher.*
For each teacher, 4 classes at 25 students each is 100 student-classes per teacher. 1600 students at 5 classes each is a total of 8000 student-classes. The number of teachers in Euclid City School is 8000 ÷ 100 = 80.

3D. <u>METHOD 1:</u>   *Strategy: Add the tens digits and units digits separately.*
After noting that $0 + 1 + 2 + 3 + ... + 8 + 9 = 45$, observe that the sum of the units digits <u>across each row</u> is $0 + 1 + 2 + 3 + ... + 8 + 9 = 45$. There are 10 such lines; hence, the sum of the units digits is $10 × 45 = 450$.

00	01	02	...	09
10	11	12	...	19
20	21	22	...	29
⋮	⋮	⋮	⋱	⋮
90	91	92	...	99
100				

Next, observe that the sum of all the tens digits <u>down each column</u> is also 45. There are 10 such columns; therefore the sum of the tens digits is also $10 × 45 = 450$.

The total sum of all the digits is $450 + 450 + 1 = 901$.
The sum of all the digits is 901.

<u>METHOD 2:</u>   *Strategy: Add all the digits in each decade.*
In each decade, the tens digit appears 10 times and the units digits are 0 through 9 inclusive. The sum of the digits for that decade, then, is $10 ×$ (tens digit) plus 45, where 45 is the sum of the units digits are 0 through 9 inclusive. There are 10 such decades.

NUMBERS	DIGIT SUM
0 - 9	$10 × 0 + 45 = 45$
10 - 19	$10 × 1 + 45 = 55$
20 - 29	$10 × 2 + 45 = 65$
30 - 39	$10 × 3 + 45 = 75$
⋮	⋮
90 - 99	$10 × 9 + 45 = 135$
A pattern has appeared!	
100	$1 + 0 + 0$ 1

The table at the right indicates the sum for each decade. Thus the sum of all the digits is $45 + 55 + 65 + ... + 135 + 1$. Adding, $(45 + 135) + (55 + 125) + (65 + 115) + (75 + 105) + (85 + 95) + 1 = 901$.

FOLLOW-UPS: For the integers from 1 to 100 inclusive: (1) How many digits are there? [192] (2) How many times does the digit 5 appear? [20]

SOLUTIONS: DIVISION M

3E. _Strategy: Perform each division from the bottom up._
One definition for $\frac{3}{4}$ is $3 \div 4$. The rule for division of fractions is "invert and multiply."
To start: $4 + \frac{3}{4}$ becomes $\frac{19}{4}$, and then $\frac{3}{\frac{19}{4}} = 3 \div \frac{19}{4} = 3 \times \frac{4}{19} = \frac{12}{19}$.

Continuing the process yields the completed result: $\dfrac{3}{4 + \dfrac{3}{4 + \dfrac{3}{4}}} = \dfrac{3}{4 + \dfrac{12}{19}} = \dfrac{3}{\dfrac{88}{19}} = 3 \times \dfrac{19}{88} = \dfrac{57}{88}$

The fraction in lowest terms equals $\frac{57}{88}$.

4A. **METHOD 1:** _Strategy: Consider the total number of brothers._
The total number of brothers is: Jakob + 3 + number of sisters. Then the family has four more brothers than sisters. Sari has 1 less sister than the total number of sisters, because Sari was already counted as one sister. Therefore, **Sari has** $4 + 1 = $ **5 more brothers than sisters.**

METHOD 2: _Strategy: Assume Jakob has various numbers of sisters, then find a pattern._
Suppose Jakob has 2, 3, 4, or 5 sisters. Then he has 5, 6, 7, or 8 brothers. Sari therefore has 1, 2, 3, or 4 sisters and 6, 7, 8, or 9 brothers. In all cases, Sari has 5 more brothers than sisters. It seems reasonable that this last statement would be true regardless of the number of sisters assigned to Jakob.

METHOD 3: _Strategy: Use algebra._
Let $M = $ number of brothers and let $F = $ number of sisters. There are $4 + F$ males, and Sari has $F - 1$ sisters. We subtract $4 + F - (F - 1) = 5$. Sari has 5 more brothers than sisters.

4B. **METHOD 1:** _Strategy: Work backwards._
Each time Ryan borrows Noelle's calculator, Noelle triples his money and then takes back 27 cents. In a "backwards" approach, we must "undo" this process. To find the number of cents Ryan had at the start of each borrowing, we must first **add back** the 27 cents and then **divide** this amount by 3. He had 0 cents after the third borrow, so:
Ryan had $(0 + 27) \div 3 = 9$ cents after the second borrow.
Ryan had $(9 + 27) \div 3 = 12$ cents after the first borrow, and therefore
Ryan had $(12 + 27) \div 3 = 13$ cents originally. **Ryan started with 13 cents.**

METHOD 2: _Strategy: Work forwards using algebra._
Suppose Ryan had N cents originally. He borrows the calculator for the first time, leaving $3N - 27$ cents. The second borrowing leaves $3(3N - 27) - 27 = 9N - 108$ cents. Finally, the third borrowing leaves $3(9N - 108) - 27 = 27N - 351$ cents. Then $27N - 351 = 0$ because he is now broke. Solving $27N - 351 = 0$ yields $N = 13$. He started with 13 cents.

**FOLLOW-UP:** Solve $5\{4[3(2x - 3) - 4] - 5\} - 6 = 69$. [3]

SOLUTIONS: DIVISION M

4C. **METHOD 1:** *Strategy: Use reasoning.*
22 students play either chess or tennis, since 3 play neither. Because $11 + 15 = 26$, then 4 students were counted twice. This could only happen if these 4 students play both chess and tennis. Thus, 4 of the 11 students who play chess also play tennis. Therefore, $11 - 4 = 7$ **students play chess, but not tennis**.

METHOD 2: *Strategy: Draw a Venn diagram.*
Same reasoning as above, but with "pictures."
Seven students play chess, but not tennis.

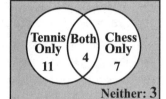

4D. **METHOD 1:** *Strategy: Consider the type of number that has an odd number of factors.*
The three factors are: the number itself (call it N), the number 1, and some third number X. Factors usually exist in pairs of different numbers: for example, the factor pairs of 12 are 1 and 12, 2 and 6, 3 and 4. A number can only have an odd number of factors if it is a perfect square: for example, the factors of 16 are 1 and 16, 2 and 8, 4 and 4 (which only counts once). Thus, N is a perfect square integer less than or equal to 150 and X is its square root.

Further, since there is no fourth factor, the only factors of X are 1 and X. Therefore, X is prime and less than 13. This yields the list: $2^2 = 4$, $3^2 = 9$, $5^2 = 25$, $7^2 = 49$, and $11^2 = 121$. **There are 5 whole numbers less than 150 with exactly three factors.**

METHOD 2: *Strategy: Investigate factors of some test numbers. Look for a pattern.*
Consider the number of factors of the first several integers. The number 1 has one factor. The numbers 2 and 3 each have two factors. The number 4 has 3 factors, and is one solution. 5 and 7 each have two factors, and 6 and 8 each have four factors. However, 9 has three factors and is another solution.

Apparently, it is perfect squares that have three factors. Test each remaining square less than 150 for the number of factors. Only 25, 49 and 121 are solutions. While this method gives us the five solutions, it does not guarantee that these are the only solutions.

FOLLOW-UP: How many whole numbers less than 25 have exactly four factors? [7]

4E. *Strategy: Sketch the path of the center of the rolling circle, then break it down into its parts.*
The path is composed of 4 straight line segments plus 4 rounded "corners". Each straight line segment has length 4. Hence $4 \times 4 = 16$. The corners of the path are composed of 4 quarter-circle arcs, each of radius 1, with a total length of $4 \times \frac{1}{4} \times 2\pi(1) = 2\pi$. Since $\pi \approx 3.14$, the total path length is $16 + 2\pi \approx 16 + 2 \times 3.14 \approx 22.28$ inches. **The center of the circle travels 22.28 inches.**

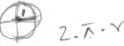

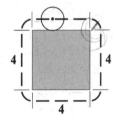

5A. **METHOD 1:**   *Strategy: Count the number of "left-out" pairs.*
Selecting three points as vertices from the five given points is the same as selecting two points to "omit". For each of the five points we can omit first, there are four other points to omit next. Note that we must divide by 2, because *AB* and *BA*, for example, count as *one* pair of left-out vertices, not as two pairs. Then the number of ways of omitting two points is $5 \times 4 \div 2 = 10$ ways. Thus the number of ways of selecting three points as vertices of a triangle is also 10, and **a total of 10 triangles are formed.**

METHOD 2:   *Strategy: Make a tree diagram for all sets of vertices.*
Use alphabetical order to ensure an accurate count.

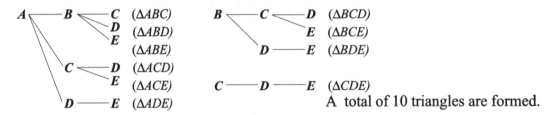

A total of 10 triangles are formed.

METHOD 3:   *Strategy: Make a list of the pairs of vertices omitted.*

AB	BC	CD	DE
AC	BD	CE	
AD	BE		
AE			

Again, 10 triangles are formed.

5B. **METHOD 1:**   *Strategy: Assume all received as few votes as possible.*
The total score for *B*, *C*, *D*, and *E* is 45 votes and each gets at least 6 votes. Since we seek the least total for *B* and no two candidates received the same number of votes, assume *E* got 6 votes, *D* got 7, *C* got 8, and *B* got 9 votes. This totals 30 votes for *E* through *B*, so we must account for 15 more votes. Add 3 to the assumed total for *E*, and 4 to the assumed totals for *D*, *C* and *B*. Then *E* got 9 votes, *D* got 11, *C* got 12, and *B* got 13. Thus, *B* received at least 13 votes.

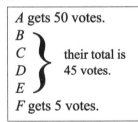

A gets 50 votes.
B
C } their total is
D } 45 votes.
E
F gets 5 votes.

METHOD 2:   *Strategy: Assume all four middle candidates received close to the average.*
The total for *B*, *C*, *D*, and *E* is 45 votes and their average is 11.25. *B*'s vote count is least when the sum of the votes for *C*, *D*, and *E* is largest. Suppose *B*, *C*, *D*, and *E* got $12 + 12 + 11 + 10 = 45$ votes. This violates rule (5). However, if *E* got one vote less, 9, then *B* got one vote more, 13. *B* received at least 13 votes.

FOLLOW-UP: What is the greatest number of votes B could have received? [24]

5C. **METHOD 1:** *Strategy: Test for divisibility by both 9 and 11.*
Any number divisible by 99 is also divisible by both 9 and 11.

If any number is divisible by 9, then the sum of its digits is also divisible by 9. Hence, the sum of its digits of 23AB3, which is 8 + A + B is a multiple of 9, namely 9 or 18; no other multiples are possible. Then $A + B = 1$ or $A + B = 10$.

If any number is divisible by 11, then the *difference* between the sum of the digits in the even places and the odd places is a multiple of 11, which includes 0. Hence, (5 + A) – (3 + B) = 2 + A – B is a multiple of 11, namely 0 or 11; no greater multiples are possible. Then $A – B = –2$ or $A – B = 9$ (not possible because $A + B = 1$ or 10). Thus, $A – B = –2$, which tells us that B is 2 more than A. It also tells us that $A + B$ can equal 10, but not 1.

Since B is 2 more than A and their sum is 10, $A = 4$ and $B = 6$. **The two-digit number AB is 46.**

METHOD 2: *Strategy: Find the quotient, digit by digit.*
(1) If 23AB3 is divided by 100, the quotient would be a three-digit number 23A, with a remainder of B3. Then if 23AB3 is divided by 99, the quotient is 23__, a three-digit number.

(2) Next, examine the units digits of 99 × 23__ = 23AB3. Because 9 × __ ends in 3, the missing units digit is 7 and 23AB3 = 99 × 237.

(3) Then 237 × 99 = 23700 – 237 = 23**46**3. The two-digit number AB is 46.

FOLLOW-UPS: *(1) If the four-digit number A17B is a multiple of 72, find A and B. [A = 1, B = 6] (2) Find both values of A such that 33A9 leaves a remainder of 7 upon division by 12. [1, 7]*

5D. **METHOD 1:** *Strategy: Create a unit of "chef-minutes".*
"Chef-minutes" are the total time required by all the chefs to do a job. To prepare 20 desserts requires 4 × 10 = 40 chef-minutes. Then each dessert requires 40 ÷ 20 = 2 chef-minutes. To prepare 75 desserts requires 75 × 2 = 150 chef-minutes. Hence, in 15 minutes, 150 chef-minutes ÷ 15 minutes per dessert = **10 chefs are required.**

METHOD 2: *Strategy: Make a chart. Adjust first for time, then for number of desserts.*
The number of desserts increases in proportion to the number of minutes or chefs. Row 1 of the table shows the given information. Row 2 changes the 10 minutes into 15 minutes by multiplying the numbers of desserts and minutes each by 1.5. Row 3 changes 30 desserts into 75 desserts by multiplying the numbers of desserts and chefs each by 2.5. 10 chefs are required.

Chefs	Desserts	Minutes
4	20	10
4	20 × 1.5 = 30	10 × 1.5 = 15
2.5 × 4 = 10	2.5 × 30 = 75	15

METHOD 3: *Strategy: Begin by finding a "unit of work" for each chef.*
Divide the given by 10: each of the 4 chefs requires 1 minute to make 2 desserts. Next multiply by 15: each of the 4 chefs requires 15 minutes to make 30 desserts. Then, to make 75 desserts, multiply by $2\frac{1}{2}$: 10 chefs are required.

5E. *Strategy: Break the figure into separate regions and add their areas.*
The region is composed of three regions. The first is three-quarters of a circle of radius 8 m and the other two regions are each quarter-circles of radius of 2 m. The total area over which the pet can roam is:

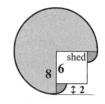

$(\frac{3}{4} \times \pi \times 8 \times 8) + 2 \times (\frac{1}{4} \times \pi \times 2 \times 2) = 48\pi + 2\pi = 50\pi$ sq m.

The whole number N is 50.

SET 12 ◆ ◆ ◆ Olympiad 1

1A. *Strategy: Use the Process of Elimination.*
The number is divisible by 5, so the ones digit is 5. There are now 6 possibilities. The second sentence eliminates _175, 1_75, and 7_15, leaving only 71_5, _715, and 17_5. The third sentence eliminates 7135 and 1735, leaving only **3715**. Considering the clues in a different order will vary details, but not the strategy nor the answer.

FOLLOW-UP: Alicia, Bill, Carl, and Dawn stand in line in some order. Carl is next to Alicia, but not next to Dawn. Alicia is ahead of Dawn, but behind Bill. In what order do they stand? [Bill, Carl, Alicia, Dawn]

1B. **METHOD 1:** *Strategy: Compare the areas of the small and large squares.*
If the small congruent squares were used like floor tiles to partition the large square *ABCD*, nine tiles would be needed. Since the area of the small square is 4 sq cm, the area of square *ABCD* is 36 sq cm. Since $\overline{AC}$ divides square *ABCD* into two triangles of equal area, **the area of triangle *ABC* is 18 sq cm.**

METHOD 2: *Strategy: Find the lengths of line segments.*
The small square has an area of 4 sq cm and a side-length of 2 cm. Then $AD = AB = BC = 6$ cm and the area of triangle *ABC* is $\frac{1}{2}(6)(6) = 18$ sq cm.

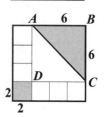

FOLLOW-UP: Method 1 can lead to an investigation of the ratio of areas of similar polygons. Consider first the shaded square and ABCD. Then consider half the shaded square and △ABC. Then consider a number of other similar polygons.
[Rule: the ratio of areas of two similar polygons equals the square of the ratio of two corresponding sides.]

1C. *Strategy: Evaluate both expressions separately.*
The value of 4 ✪ 5 is $4 + 4 - 5 = 3$. Since 4 ✪ 5 has the same value as 6 ✪ ☐, 6 ✪ ☐ = 3. Evaluating: $6 + 6 - ☐ = 3$, or $12 - ☐ = 3$. Thus, **the value of ☐ is 9.**

FOLLOW-UP: Assorted "Guess My Rule" activities are appropriate. For example: If 2 ◈ 6 = 6, 3 ◈ 6 = 6, 4 ◈ 6 = 12, 5 ◈ 6 = 30, and 9 ◈ 6 = 18, find the value of 10 ◈ 6. [30, if ◈ means LCM]

SOLUTIONS: DIVISION M

1D. **METHOD 1:** *Strategy: Use the definition of "average".*
An overall average is the total of all scores divided by the number of scores. For all five tests the average of Carlos' grades is 84, so their sum is $5 \times 84 = 420$. For his first three tests the average of his grades is 80, so their sum is $3 \times 80 = 240$. For his last two tests the sum must be $420 - 240 = 180$, so **their average is** $180 \div 2 = 90.$

METHOD 2: *Strategy: "Balance" the deviations from the average.*
The sum of the differences between the average and each of the grades must be zero. The overall average of all five grades is 84, but the average of just the first three grades is 80. Thus, each of the first three grades, on average, is 4 points less than the overall average of 84, for a total "deficit" of $3 \times 4 = 12$ points. To attain an overall average of 84, the average of the last two grades must be $12 \div 2 = 6$ points more than 84. Then the average of Carlos' last two grades is $84 + 6 = 90$. Think: $(84 - 4), (84 - 4), (84 - 4), (84 + \_\_)$, and $(84 + \_\_)$.

FOLLOW-UPS: (1) Does an average of 80 mean that each of his first 3 tests is 80? (2) A car travels at a steady rate of 40 mph for 3 hours. Then it returns along the same route in 2 hours. What was its average rate on the return trip and what is its overall average rate? [60 mph, 48 mph] (3) A car travels 15 miles at an average rate of 30 mph. At what rate must it travel on the return trip in order to average 60 mph for the whole trip? [It is not possible!]

1E. **METHOD 1:** *Strategy: Find how much money she lost, not how much she made.*
By leaving 2 weeks early, Melanie lost $600 - 150 = 450$. This is $225 per week. At this rate, in eight weeks she would have earned a total of $1800. **The value of the computer is** $1800 - 600 = \mathbf{\$1200.}$

METHOD 2: *Strategy: Use algebra.*
Let x = the value of the computer in dollars. Then $x + 600 =$ her pay for 8 weeks, in dollars and $x + 150 =$ her pay for 6 weeks, in dollars. She worked for $\frac{3}{4}$ of the agreed time. It follows that $x + 150 = \frac{3}{4}$ of $(x + 600)$. Then:

$$x + 150 = \tfrac{3}{4}(x + 600)$$

Distribute the $\frac{3}{4}$ to each term:	$x + 150 = \frac{3}{4}(x) + \frac{3}{4}(600)$
Simplify: $\frac{3}{4}(600)$:	$x + 150 = \frac{3}{4}x + 450$
Subtract 150 from each side:	$1x \quad\quad = \frac{3}{4}x + 300$
Subtract $\frac{3}{4}x$ from each side:	$\frac{1}{4}x \quad\quad = \quad\quad 300$
Multiply both sides by 4:	$x \quad\quad = \quad\quad 1200$

The value of the computer is $1200.

(Note: On these contests, algebraic methods are shown as alternate solutions.)

FOLLOW-UP: Sonia has a total of 70 animals, all cows or chickens. Her animals have a total of 200 legs. How many cows does Sonia have? [30 cows]

SOLUTIONS: DIVISION M

2A. _Strategy: Find the factor pairs of 96._
Factoring 96 in pairs yields 1×96, 2×48, 3×32, 4×24, 6×16, and 8×12. There are only 12 months and between 28 and 31 days, inclusive, in a month. The only interesting dates in 1996 were **April 24** (1/24/96), **June 16** (6/16/96), **August 12** (8/12/96), and **December 8** (12/8/96).

FOLLOW-UPS: (1) List all interesting dates for 1990 [3/30, 5/18, 6/15, 9/10, 10/9] _(2) Since 1980, which years have the most interesting dates?_ [1984 and 1990 – 5 interesting dates each.]

2B. _Strategy: Work from right to left._
(1) $E \times 3$ ends in 1. Then $E = 7$ and 2 is carried. (2) Since 2 more than $D \times 3$ ends in 7, $D \times 3$ ends in 5. Then $D = 5$ and 1 is carried. (3) Since 1 more than $C \times 3$ ends in 5, $C \times 3$ ends in 4. Then $C = 8$ and 2 is carried. (4) Since 2 more than $B \times 3$ ends in 8, $B \times 3$ ends in 6. Then $B = 2$ without carry. **The value of B is 2.** (Check: Since $A \times 3$ ends in 2, $A = 4$ and 1 is carried. Then $1 \times 3 + 1 = 4$. Thus the multiplicand is 142857, the product is 428571, and $142857 \times 3 = 428571$.)

```
1 A B C D E
×         3
A B C D E 1
```

FOLLOW-UP: Express the decimal equivalents of $\frac{1}{7}$, $\frac{2}{7}$, $\frac{3}{7}$, $\frac{4}{7}$, $\frac{5}{7}$, $\frac{6}{7}$ as repeating decimals.

2C. **METHOD 1:**   _Strategy: Subtract the areas of the two squares and divide by 4._
The area of the larger square is $14^2 = 196$ sq cm. The area of the smaller square is $8^2 = 64$ sq cm. The sum of the areas of the four congruent rectangles is $196 - 64 = 132$ sq cm. Then **the area of the shaded rectangle is** $132 \div 4 =$ **33 sq cm.**

METHOD 2:   _Strategy: Find the width and length of the shaded rectangle._
The width of the shaded rectangle $(14 - 8) \div 2 = 3$ cm. Then the length of the shaded rectangle is 14 cm $- 3$ cm $= 11$ cm. The area of the shaded rectangle is $11 \times 3 = 33$ sq cm.

FOLLOW-UP: Suppose the area of each of the congruent rectangles in the diagram is 15 sq cm and all measures are whole numbers. What are the areas of the inner and outer squares? [The answer consists of two possibilities: 256 and 196, or 64 and 4.]

2D. _Strategy: Eliminate the impossible (Process of Elimination)._
The least number of dollars possible is $7 \times \$4 = \$28$. A sales total of $26 is impossible.
The greatest number of dollars possible is $7 \times \$10 = \$70$. A sales total of $75 is impossible.
The price of each ticket is even. Any odd sales total ($37, $57) is impossible.

This leaves $48 and $68 as perhaps possible. For seven $10 tickets the sales total is $70. However, for six $10 tickets and one $6 ticket the sales total is $66. Additional replacements would further reduce the sales total. Thus, a sales total of $68 is impossible. **The only possible total sales figure is $48.** The sum of $48 can be accomplished two ways:
$10 + 10 + 10 + 6 + 4 + 4 + 4$, or $10 + 10 + 6 + 6 + 6 + 6 + 4$.

FOLLOW-UPS: *(1) Are there any other values from $28 to $70 which are impossible? [Only the odd values.] (2) Can you construct a similar question if the tickets are $4, $8, and $10 each? (3) If each of 6 tickets cost $3, $5, or $9, what sums are impossible? [All numbers except 54 and 18, 20, 22, ... , 50]*

2E. **METHOD 1:** *Strategy: Find the prime factors of 1000.*
Any number N that ends in 3 zeros is a multiple of 1000. The prime factorization of 1000 is $5 \times 5 \times 5 \times 2 \times 2 \times 2$. Any factor of 1000 is also a factor of N. The three least whole numbers for which 5 is a factor are 5, 10, and 15. Then the minimum value of N is 15 and N is less than 20. The three least whole numbers for which 2 is a factor are 2, 4, and 6. Then N is at least 6. (Actually, 2 is a factor of $2 \times 4 \times 6$ *four* times!) **The least value of N for which $N!$ contains three factors of 5 and at least 3 factors of 2 is 15.**

METHOD 2: *Strategy: Examine the first several integers for factors of 5 and 2.*
Any number ending in 3 zeros must have the factor pair (5×2) appear 3 times because $10 = 5 \cdot 2$. Showing prime factorization, list the first several consecutive numbers and check for the factor pair 5×2:

1	4 = 2 · ②	7	10 = ⑤ · ②	13
2	5 = ⑤	8 = 2 · 2 · 2	11	14 = ② · 7
3	6 = 3 · 2	9 = 3 · 3	12 = 3 · 2 · 2	15 = 3 · ⑤
	Pair #1		Pair #2	Pair #3

The third factor pair (5×2) first appears when $N = 15$.

FOLLOW-UPS: *(1) What is the greatest value of N for which N! ends in exactly 3 zeros? [19] (2) How many terminal zeros are in the products of 20!, 25!, and 30! ? [4, 6, 7] (3) Why is it impossible for N! to have 5 terminal zeros?[Every 5th multiple of 5 has an additional factor of 5] (4) If 2^n is a factor of 30!, what is the greatest value that **n** can have? [26]*

SET 12   **Olympiad 3**

3A. *Strategy: Count in groups of similar angles.*
Ten acute angles can be found:
1) $\angle B$, $\angle ADE$, $\angle AED$, $\angle C$;
2) $\angle BAD$, $\angle DAE$, $\angle EAC$;
3) $\angle BAE$, $\angle CAD$; and
4) $\angle BAC$.

3B. _Strategy: Consider one condition at a time._

According to the second condition, the least possibility that produces two digits is 3×4 and the greatest possibility is 8×9. Then **the numbers are 12** $= 3 \times 4$ **and 56** $= 7 \times 8$, each of which satisfy both conditions. In the product 90 (from 9×10), 9 and 0 are not consecutive integers.

3C. _Strategy: Solve for the reciprocal of N and then for N itself._

$$\frac{1}{5} = \frac{1}{6} + \frac{1}{N}$$
$$\frac{1}{5} - \frac{1}{6} = \frac{1}{N}$$
$$\frac{1}{30} = \frac{1}{N}$$
$$\frac{1}{N} = \frac{1}{30} \quad \textbf{The value of } N \textbf{ is 30.}$$

FOLLOW-UP: _Solve for each N and then find a pattern:_

$$\frac{1}{3} = \frac{1}{4} + \frac{1}{N} \qquad\qquad \frac{1}{4} = \frac{1}{5} + \frac{1}{N} \qquad\qquad \frac{1}{99} = \frac{1}{100} + \frac{1}{N} \quad [12;20;9900]$$

3D. _Strategy: Calculate the number of minutes to be lost._

To be again correct, the clock must lose a total of 12 hours, which equals 720 minutes. At 12 minutes lost per hour, **60 hours must elapse.**

(Note: _The question asks for hours. Answers such as_ $2\frac{1}{2}$ _days or 2 days, 12 hours are not correct.)_

FOLLOW-UPS: _(1) What is the correct time when the clock first shows 5:20?_ [5:55] _(2) At the next correct time of 9:30, what time does the clock show?_ [8:12]

3E. _Strategy: Count paths to each letter separately._

1) From O there are 2 paths to L.

2) From each L there are 2 paths that lead to a Y (Think L_1Y_1 and L_1Y_2). Then from O to Y there is a total of $2 \times 2 = 4$ paths.

3) From each Y there is one path to M. Then from O to M there is a total of $4 \times 1 = 4$ paths.

4) From each M there are 2 paths that lead to a P. Then from O to P there is a total of $4 \times 2 = 8$ paths.

5) From each P there are 3 paths that lead to an I. Then from O to I there is a total of $8 \times 3 = 24$ paths.

6) From each I there are 2 paths that lead to an A. Then from O to A there is a total of $24 \times 2 = 48$ paths.

7) From each A there is only 1 path that leads to the D. Thus from O to D **there is a total of** $48 \times 1 = \textbf{48 paths.}$

```
      O
   L     L
   Y     Y
      M
   P     P
 I    I    I
   A     A
      D
```

FOLLOW-UPS: _(1) Moving only up or to the right along the lines shown, how many paths can be traveled from A to B?_ [70] _(2) Research Pascal's Triangle._

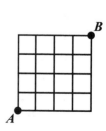

4A. **METHOD 1:** *Strategy: Work backwards.*
Represent the sequence as $\boxed{D} \xrightarrow{\times 4} \boxed{C} \xrightarrow{+8} \boxed{B} \xrightarrow{\div 4} \boxed{A} \xrightarrow{-8} \boxed{4}$. The end result is 4, so $A = 12$. Then $B = 48$, $C = 40$ and $D = 10$. **Carol chose the number 10.**

METHOD 2: *Strategy: Use algebra.*
Represent the starting number as n.
Carol multiplies n by 4 (result: $4n$),
then adds 8 (result: $4n + 8$),
then divides by 4 (result: $n + 2$), and
finally subtracts 8 (result: $n - 6$).
We are told that this result equals 4. That is, $n - 6 = 4$, so $n = 10$.
Carol chose the number 10.
(Note: Algebraic solutions are shown as alternatives to more elemental approaches.)

4B. *Strategy: Eliminate all factors of 2.*
Repeated division by 2 leads to $4664 = 2 \times 2332 = 2 \times 2 \times 1166 = 2 \times 2 \times 2 \times 583$. **Thus, the greatest odd factor of 4664 is 583.**

FOLLOW-UPS: (1) List all odd factors of 4664 [11, 53, 583 – this reviews the test for divisibility by 11.]
(2) List all odd factors of 3360. [1, 3, 5, 7, 15, 21, 35, 105] *How do you know when you have found them all?* [Use a factor table.]

4C. **METHOD 1:** *Strategy: Break up the table into 2 tables.*
Move the even-numbered rows into a second table, to the right of the original table, and reverse the columns in the second table. Each entry is now 12 less than the entry directly below it. Then 100 is in the same column as 88, 76, 64, and so on. The number at the top of the column is the remainder if 100 is divided by 12, namely 4. Since column F contains the number 4, **the number 100 appears under the letter F.**

A	B	C	D	E	F	F	E	D	C	B	A
		1	2	3	4	5	6	7	8	9	10
11	12	13	14	15	16	17	18	19	20	21	22
23	24	25	26	27	28	29	30	31	32	33	34
35	36	37	38	39	40	41	42	43	44	45	46
											... and so on

METHOD 2: *Strategy: Find a pattern in a column.*

Pick any column, say, the original column *C*. Reading down, the differences from one row to the next are consistent. In column *C*, they are 7, 5, 7, 5, 7, 5, ... Then column *C* contains 1, 8, 13, 20, 25, 32, 37, 44, 49, 56, 61, 68, 73, 80, 85, 92, and 97. Thus, the number 100 appears three columns to the right of column *C*, under the letter *F*.

FOLLOW-UPS: *(1) Under which letter would the year 2000 appear? [C] (2a) Suppose 1 is a first-row entry in a different column. If the entry 250 is under the letter B, under which two letters could the entry 1 appear? [B or E] (2b) Suppose 1 is a first-row entry in a different column from the one in part (2a). Is it possible for the entry 250 to be under the letter E? [no]*

4D. *Strategy: Use the test of divisibility for 9 first.*

The number 3*D*8 is divisible by 9 so the sum of the digits, $11 + D$, is divisible by 9. Then $D = 7$. Since $ABC - 378 = 269$, $ABC = 378 + 269$, which equals 647. **Then *ABC* = 647.**

4E. **METHOD 1:** *Strategy: Build up from simpler cases.*

Two chords can intersect in 1 point.

A third chord can intersect the first two chords in 2 more points.

A fourth chord can intersect the first three chords in 3 more points.

A fifth chord can intersect the first four chords in 4 more points.

A sixth chord can intersect the first five chords in 5 more points.

The greatest number of intersections is $1 + 2 + 3 + 4 + 5 = \mathbf{15}$.

METHOD 2: *Strategy: Use Combinatorics.*

Each of the six chords can intersect any of the 5 others. That would seem to indicate 30 intersections. However, chord #1 intersects chord #2 in the same point that chord #2 intersects chord #1. Thus, the greatest number of intersections is $(6 \times 5) \div 2 = 15$.

FOLLOW-UPS: *(1) Upon entering a room, ten people shake hands with each other once. How many handshakes took place? [45] (2) Each of the 16 teams in a tournament plays every other team once. How many matches must be scheduled? [120] (3) Is there an easy way to calculate the maximum number of intersections if there are 100 chords? [(100 × 99) ÷ 2]*

SET 12   **Olympiad 5**

5A. **METHOD 1:** *Strategy: Partition the set according to =, >, and <.*

Nine counting numbers less than 100 have equal digits: 11, 22, 33, ..., 99. This leaves 90 one- or two-digit numbers whose digits are not equal. Of these 90, 45 have their tens digit greater than the ones digit and the other 45 have their tens digit less than the ones digit. **The required set contains 45 numbers.**

METHOD 2:   *Strategy: Partition the set of numbers by tens.*

List of Numbers	Quantity
10	1
20 21	2
30 31 32	3
⋮ ⋮ ⋮ ⋱	⋮
90 91 92 ... 98	9
TOTAL =	45

The required set contains 45 numbers, as shown.

FOLLOW-UPS: (1) How many two-digit counting numbers have their tens digit less than their ones digit? [45] (2) For how many three-digit natural numbers are the digits in descending order? [120] (3) How is problem 5A related to problem 4E? [Both involve triangular numbers.]

5B. *Strategy: Find the dimensions of each rectangle.*

The area of a rectangle equals the product of its length and width. Then $a \times c = 35$ and $a \times d = 10$. The value of a is either 5 or 1. But a cannot be 1, or else c would be 35 and 42 would have to be a multiple of 35. Thus, $a = 5$. Then $c = 7$, $d = 2$ and $b = 6$.
It follows that $N = b \times d = 6 \times 2$. **The value of N is 12.**

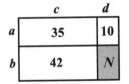

FOLLOW-UPS: (1) Research the medieval method of lattice multiplication. (2) Use the upper rectangles to demonstrate the distributive principle: $a(c + d) = ac + ad$. In this case, $5 \times 7 + 5 \times 2 = 5 \times 9$; that is, the sum of the areas of the small rectangles (35 + 10) equals the area of the large upper rectangle ($5 \times 9 = 45$).

5C *Strategy: Subtract corresponding terms to get a constant difference.*

To maximize Karen's total, choose the nine greatest values available. Then Rob's total must come from the nine least values available. Subtract each from its counterpart so as to get the common difference. Enter the given values as adjustments to the totals.

$$\text{Karen's total} = (81 + 82 + 83 + \ldots + 90) - 89 + 74$$
$$\underline{\text{Rob's total} = (71 + 72 + 73 + \ldots + 80) + 89 - 74}$$
$$\text{Karen's total} = (81 + 82 + 83 + \ldots + 90) - 15$$
$$\underline{\text{Rob's total} = (71 + 72 + 73 + \ldots + 80) + 15}$$

Subtracting, $(10 + 10 + 10 + \ldots + 10) - 30$
Difference = (100) -30
Difference = 70   **Karen's total is 70 more than Rob's total.**

(Note: Most students may solve this problem by brute force, but this approach offers them a way to avoid heavy computation. Many problems allow us to use the character of the numbers to ease the arithmetic.)

FOLLOW-UP: Suppose Karen and Rob each selects 1000 slips, which are numbered from 1 to 2000. If Karen selects the "45" and Rob selects the "1945", what is the greatest amount by which Karen's total can exceed Rob's total? [996,200]

5D. _Strategy: Find the cost of the gifts and then the number of members._
The members originally paid $90 + $60 = $150 for the party. Buying gifts cost $210 − $150 = $60. At $2 each, a total of 30 members participate. Since the 30 members at first raised $150, each originally paid $5. To raise $90 at $5 each, **there must be 18 girls.**

5E. _Strategy: Use the fact that all 3 digits of AAA are equal._
AAA is a multiple of 111. Factor 111: 37 × 3 = 111. Since _AB_ × _C_ = _AAA_, _AB_ = 37 and _AAA_ = 333. Then _C_ = 333 ÷ 37 = 9. **The value of _C_ is 9.**

SET 13 ◆ ◆ ◆ Olympiad 1

1A. _Strategy: Find the remainder after division by 12._
Every 12 months from November will be another November. Because 96 is a multiple of 12, 96 months from now will be November. **The 100ᵗʰ month**, 4 months later, **will be March.**

**FOLLOW-UPS:** (1) What day of the week occurs 1000 days from the day before yesterday? [Yesterday] (2) A school is on a 6-day cycle. If the school year contains 175 class days, which day of the cycle is the last day of school? [Day 1]

1B. **METHOD 1:** _Strategy: Compare the sum of some parts to the whole amount._
Erin and Ari, Cara and Dara have a total of 20 + 30 = $50. Thus **Barry has** 85 − 50 = **$35.**

METHOD 2: _Strategy: Compare the whole amount to the sum of all parts._
Ari and Barry, Cara and Dara, and Erin and Ari have a total of 40 + 30 + 20 = $90. Since everyone is counted once except Ari, who is counted twice, Ari has 90 − 85 = $5. Then Barry has $35.
(Note: Algebra can also be used to solve this problem, following closely one of the above approaches.)

1C. _Strategy for all three methods: Use a Venn Diagram._
METHOD 1: Since 10% neither own computers nor are in band, 90% either own computers, are in band, or both. However, the sum of 80% and 40% is 120%, but only 90% are in at least one category. Therefore, 30% must have been counted twice. Thus, **30% of all the mathletes both own computers and are in band.**

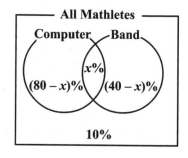

METHOD 2: Label the regions 80 − _x_, 40 − _x_, and _x_, as shown. Then (80 − _x_) + (_x_) + (40 − _x_) = 90. Solving _x_ = 30.

METHOD 3: In percentage problems it often is helpful to let the total number be 100, so assume that there are exactly 100 mathletes. Then follow one of the previous methods.

1D. _Strategy: Extend the process of cancellation._
Cancel identical numerators and denominators with each other (that is, divide out each common factor greater than 1). This can be done eleven times. Then we are left with $\frac{3}{27}$. **In lowest terms, the product is $\frac{1}{9}$.**

FOLLOW-UPS: _(1) What is the product of $1\frac{1}{2} \times 1\frac{1}{3} \times 1\frac{1}{4} \times \ldots \times 1\frac{1}{25}$?_ [13] _(2) If 5! means $5\cdot4\cdot3\cdot2\cdot1$, what is the value of $\frac{8!}{2! \times 6!}$?_ [28]

1E. **METHOD 1:** _Strategy: Organize counting by the number of regions needed to make a triangle._
In the table below, the first column categorizes triangles by the number of regions each contains. The second column specifies the regions within each triangle, and the third column gives the number of such triangles.

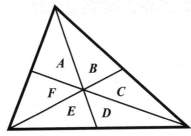

Number of Regions	Regions	Number of Triangles
1	_A, B, C, D, E, F_	6
2	_AF, ED, CB_	3
3	_AFE, BCD, FAB, EDC, DEF, CBA_	6
6	_ABCDEF_	1
TOTAL =		**16**

There are 16 triangles of all sizes.

METHOD 2: _Strategy: Organize the triangles by angles._
How many triangles contain angle 1, 2, 3, and so on? In this table, the first column categorizes triangles by one of its angles. The second column specifies the regions that form each triangle (other than those previously counted), and the third column counts the number of such triangles.

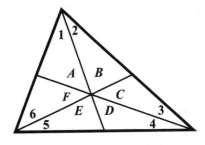

Angle	Regions	Number of Triangles
1	_A, AF, AFE_	3
2	_B, BC, BCD_	3
3	_C, CBA_	2
4	_D, DE, DEF_	3
5	_E, EDC_	2
6	_F, FAB_	2
ALL	_ABCDEF_	1
TOTAL		**16**

FOLLOW-UPS: _(1) How many squares of all sizes can be found on the standard 8 by 8 checkerboard?_ [204] _(2) How many rectangles of all sizes can be found on the standard checkerboard?_ [1296] _(3) In diagram for Problem 1E, if 3 of the 7 distinct points of intersection are chosen at random, what is the probability that these points determine a triangle?_ [$\frac{29}{35}$; hint — ignore the given line segments.]

2A. _Strategy: Try to avoid picking 3 red marbles._
To <u>guarantee</u> 3 red marbles, I must first try to avoid picking 3 red marbles. Suppose I pick all 30 marbles that are not red. Then the next 3 marbles must be red. Thus, **I must pick 33 marbles in order to guarantee that 3 marbles are red.**

FOLLOW-UPS: (1) How many marbles must I pick in order to guarantee that 3 marbles are of the same color (any color)? 4 marbles, 5? 6? n? $[9; 13; 17; 21; 4n-3]$ _(2) What is the probability of picking 3 red marbles?_ $[\frac{10}{40} \times \frac{9}{39} \times \frac{8}{38} = \frac{3}{247}]$

2B. **METHOD 1:**   _Strategy: Set up a proportion._
Since Jessica won 20% of the matches, Emily won 80% of the matches. Thus the ratio of Jessica's wins to Emily's wins is 1:4. Then 1:4 = ?:12, so **Jessica won 3 times**.

METHOD 2:   _Strategy: Base a table on the 20% figure._
Since Jessica won 20% of the matches, she won 1 of every 5 matches. Suppose the total number of matches was 5, 10, 15, etc. Then the table shows the number of wins for each girl in each case. Since Emily had 12 wins, Jessica had 3 wins.

Matches Played	Matches Emily Wins	Matches Jessica Wins
5	4	1
10	8	2
15	12	3

2C. **METHOD 1:**   _Strategy: Use tests of divisibility for factors of 12._
Any number divisible by 12 is also divisible by all factors of 12. Thus $839A2$ is divisible by both 3 and 4. Since $839A2$ is divisible by 3, then $8 + 3 + 9 + A + 2 = 22 + A$ is divisible by 3. Hence $A = 2$ or 5 or 8. Since $839A2$ is divisible by 4, then the number formed by the last two digits, $A2$, is divisible by 4. Thus 22 or 52 or 82 is divisible by 4. This is true only for 52, so **A represents the digit 5.**

METHOD 2:   _Strategy: Use Long Division._
Divide as shown at the right. Since the number $B2$, obtained in the last subtraction, is divisible by 12, $B2$ represents either 12 or 72. Because $11A$ is greater than 108 by at least 2, B represents 7. Thus A represents the digit 5.

```
        6 9 9 C
    12)8 3 9 A 2
       7 2
       ─────
       1 1 9
       1 0 8
       ─────
         1 1 A
         1 0 8
         ─────
           B 2
```

FOLLOW-UPS: (1) If the five-digit number A6A41 is divisible by 9, what digit does A represent? $[8]$ _(2) If the four-digit number T37V is divisible by 88, what digit does T represent?_ $[2]$

2D. _Strategy: Use the meaning of fraction, as division._

The meaning of any fraction $\frac{a}{b}$ is $a \div b$. Therefore, $\frac{6}{.3}$ means $6 \div 0.3$, whose value is 20. Similarly, $\frac{3}{.06}$ means $0.3 \div 0.06$, whose value is 5. **The value of the sum is 25.**

Alternately, $\frac{6}{.3} = \frac{60}{3} = 20$, $\frac{3}{.06} = \frac{30}{6} = 5$, so that $\frac{6}{.3} + \frac{3}{.06} = 20 + 5 = 25$.

FOLLOW-UPS: _Find the value of each of the following:_ (1) $\dfrac{0.1}{0.1 + \dfrac{0.1}{1 + 0.1}}$ $\left[\frac{11}{21}\right]$ (2) $\dfrac{1\frac{3}{4}}{2\frac{1}{2} + 1.25}$ $\left[\frac{7}{15}\right]$

2E. **METHOD 1:** _Strategy: Work with lengths of segments._

The sides of the squares A, B, and C are 3, 4, and 3 m long, respectively. Therefore, each side of the large square is 10 m long. Then the bases of the trapezoid are 4 m and 10 m long. Partition the shaded region into a central rectangle 4×6 (area = 24) and two right triangles each of whose legs are 3 and 6 (areas = 9 and 9). **The area of the shaded region is** $24 + 9 + 9 = $ **42 sq m.**

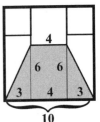

METHOD 2: _Strategy: Subtract the unshaded region from the whole figure._

The area of the large square is $10 \times 10 = 100$ sq m. If unshaded trapezoids A and B are placed together as shown (after flipping trapezoid A over), they form a rectangle of area $3 \times 8 = 24$ sq m. Then the total area of all the unshaded regions is $24 + 9 + 16 + 9 = 58$ sq m, and the area of the shaded region is $100 - 58 = 42$ sq m.

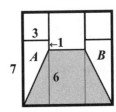 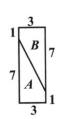

METHOD 3: _Strategy: Use an appropriate formula, if known._

As in Method 1, the bases of the trapezoid are 4 m and 10 m long. Its altitude is $10 - 4 = 6$ m. By formula, the area of the shaded region is $\frac{1}{2}(6)(4 + 10) = 42$ sq m.

3A. **METHOD 1:** _Strategy: Group._

$(1 - 3) + (5 - 7) + (9 - 11) + (13 - 15) + (17 - 19) = {}^-2 + {}^-2 + {}^-2 + {}^-2 + {}^-2 = {}^-10.$ **The result is $^-$10.**

METHOD 2: _Strategy: Group differently._

$1 + ({}^-3 + 5) + ({}^-7 + 9) + ({}^-11 + 13) + ({}^-15 + 17) - 19 = 1 + 2 + 2 + 2 + 2 - 19 = {}^-10.$

METHOD 3: _Strategy: Group positives and negatives separately._

$(1 + 5 + 9 + 13 + 17) - (3 + 7 + 11 + 15 + 17) = 45 - 55 = {}^-10.$

FOLLOW-UPS: _(1) Mr. Arcaro buys a horse for \$40, sells it for \$50, buys it back for \$60, and finally sells it again for \$70. How much did he end up making?_ [\$20] _(2) A snail falls into a well 30 feet deep. Trying to get out, each day it crawls up 3 feet, but each night it slips back 2 feet. How many days does it take the snail to reach the top?_ [28; the snail is out before it can slip back on the last day.]

3B. **METHOD 1:**   *Strategy: Compare purchases to each other.*
Because Dan buys two apples more than Chris but two bananas fewer than Chris, Dan pays $1 more. Thus, if Dan buys 1 apple more than Chris but 1 banana fewer than Chris, he would pay 50¢ more. Therefore **one apple costs 50¢ more than one banana.**

METHOD 2:   *Strategy: Use algebra.*
Suppose one apple costs A cents and one banana costs B cents. Then the equations are $5A + 3B = 570$ and $3A + 5B = 470$. Subtracting the two equations, $2A - 2B = 100$, so $A - B = 50$.

FOLLOW-UPS: *(1) How much would it cost to buy 8 apples and 8 bananas? [$10.40] One apple and one banana? [$1.30; notice that in none of the above was it necessary to know what one apple cost!] (2) Each balance scale below shows objects in perfect balance. How many □ will balance a ●? [6]*

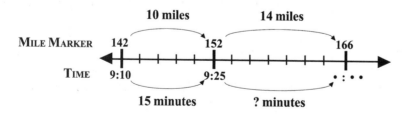

3C.   *Strategy: Begin by finding the value of E.*
Because $U + E = U$ with no carry, $E = 0$. Since $U \times M$ ends in zero and neither can be zero, then either $UM = 52$ or $UM = 25$. UM is not 52 because $U2 \times U2$ does not end in 2. Then $25 \times 25 = 625$ matches all conditions in the example and $S\ U\ M$ **represents 625.**

$$
\begin{array}{r}
2\,5 \\
\times\ \ 2\,5 \\
\hline
1\,2\,5 \\
5\,0\ \ \\
\hline
6\,2\,5
\end{array}
$$

3D. **METHOD 1:**   *Strategy: Find her rate of speed.*
Samantha travels 10 miles in 15 minutes. Her rate of speed is 2 miles every 3 minutes. Thus to travel another $2 \times 7 = 14$ miles, she needs $3 \times 7 = 21$ more minutes.
Finally, 21 minutes after 9:25 is 9:46. **Samantha passes mile marker 166 at 9:46 AM.**

METHOD 2:   *Strategy: Use proportions.*
Samantha takes x minutes to travel between markers 152 and 166. Then $10:15 = 14:x$, so $x = 21$. She passes mile marker 166 at 9:46 AM.

METHOD 3:   *Strategy: Simplify to a unit rate.*
Since 10 miles require 15 minutes to travel, each mile requires 1.5 minutes to travel. Then, to travel 14 miles, she would need $1.5 \times 14 = 21$ minutes. She passes marker 166 at 9:46 AM.

(Note: Often a diagram is helpful with any of the above.)

3E. <u>*Strategy*</u>: *Consider one vertex at a time.*
To reach point *C* from point *A* there are 3 paths: *AFC*, *AC*, and *AHC*. To reach point *D* from point *A* there are 5 paths: *AFD*, *AFGD*, and the 3 paths through point *C*. Similarly, to reach point *E* from point *A* there are also 5 paths. Then **to reach point B from point A a total of 13 paths can be traced**: 5 through point *D*, 5 through point *E*, and the 3 through point *C* that do not pass through point *D* or *E*.

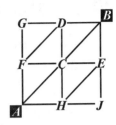

FOLLOW-UPS: *(1) How many paths are possible if the figure is a 3×3 square? [63] (2)* *EXPLORATION*: *If we extend the figure to a 5×5 square, what patterns can be discovered about the number of paths along the vertices in a single direction? (Note: For discussion purposes, students can practice the language of the coordinate system by referring to point A as (0,0) and point F as (0,1).)*

4A. <u>*Strategy*</u>: *Utilize place value.*
To obtain the largest difference, the hundreds digits of the two numbers must be 1 and 9. Then the three-digit numbers are 9A1 and 1A9. Any value of the tens digit produces the same difference after subtraction. Suppose it is 4. Then **the greatest possible difference is** 941 – 149 = **792.**

FOLLOW-UP: *Mathemagic: (Ask the whole class to do the following.) Pick a 3-digit number that is not a palindrome. Reverse the digits, adding a leading zero if it is needed to make the reversal a 3-digit number also. Subtract them. Reverse the digits on the result. Add the result and its reversal. What is your answer? [1089, regardless of the number chosen.]*

4B. **METHOD 1:**   <u>*Strategy*</u>: *Compare each remainder to its related divisor.*
Notice that in each case, the remainder is one less than the divisor. Therefore, if we add 1 to the number we seek, we get the next greater multiple of both 3 and 4 simultaneously. Therefore we are seeking one less than the least common multiple of 3 and 4 which is greater than 50. The common multiples of 3 and 4 are all the multiples of 12. The least multiple of 12 which is greater than 50 is 60, and one less than that is 59. **The least number that satisfies all conditions is 59.**

METHOD 2:   <u>*Strategy*</u>: *Make two lists and compare them.*
Consider the counting numbers greater than 50. Those that leave a remainder of 2 upon division by 3 are: 53, 56, 59, 62, 65, 68, 71, and so on. Those that leave a remainder of 3 upon division by 4 are: 51, 55, 59, 63, 67, 71, 75, and so on. The least number on both lists is 59.

FOLLOW-UPS: *(1) What is the least integer that satisfies the above and is greater than 2000? [2003] (2) What is the least counting number that has a remainder of 2 when divided by 8, 3 when divided by 9, and 4 when divided by 10? [360 – 6 = 354]*

SOLUTIONS: DIVISION M

4C. **METHOD 1A:** *Strategy: Work from the middle integer.*
As indicated in the "before-and-after" diagrams at the right, the middle term of the 7 consecutive integers is also their average. Then the middle term is $105 \div 7 = 15$ and **the sum of the least and greatest of these integers is** $2 \times 15 = 30$.

Before

After

METHOD 1B: *Strategy: Use algebra.*
The same thinking as in Method 1A can be employed algebraically by representing the *middle* integer as x and writing the equation as:
$(x-3) + (x-2) + (x-1) + (x) + (x+1) + (x+2) + (x+3) = 105$.
So, $7x = 105$, then $x = 15$, and finally $(x-3) + (x+3) = 2x = 30$.

METHOD 2: *Strategy: Start from the least integer.*
Represent the least integer as S. Then the other six integers are $S+1, S+2, S+3, S+4, S+5,$ and $S+6$. So $7S + 21 = 105$, $7S = 84$, and $S = 12$. The sum of least and greatest integers is $12 + 18 = 30$.

EXPLORATIONS: (1) How would the solution be affected if the problem specified 7 consecutive odd integers? (2) Suppose the sum of six consecutive integers is 135. How is the average related to the middle numbers? (3) How does the median of a set of consecutive integers compare to its arithmetic mean?

4D. **METHOD 1:** *Strategy: Consider what is paid, not what is saved.*
A discount of "20% off a price" is equivalent to "80% of the price". Hence, the sale price is 80% of the original price. A second subsequent discount of 20% gives a final price that is 80% of the sale price, that is 80% of 80% = 64% of the original price. **Thus, the single discount is 100% − 64% = 36%.** The same conclusion can be reached by using $\frac{4}{5}$ of $\frac{4}{5}$.

METHOD 2: *Strategy: Assign a convenient value to the original price.*
Assume that the original price was $100. Then a 20% discount yields a sale price of $100 − $20 = $80. Then, because a 20% discount on $80 is $16, the final price is $80 − $16 = $64. Thus, the total discounted amount is $36. The single discount is $36 ÷ $100 = 36%.

FOLLOW-UPS: (1) Which yields the best savings: consecutive discounts of 10% and 30%, or of 20% and 20%, or a single discount of 40%? [the single discount] (2) After a single 20% discount, what percent increase is needed to return to the original price? [25%]

4E. **METHOD 1:** *Strategy: Consider vertical and horizontal toothpicks separately.*
Observe that in each row there is one more vertical toothpick than there are squares. Thus a row that has 8 squares across would have 9 vertical toothpicks. Since the rectangle has 6 rows, there is a total of $6 \times 9 = 54$ vertical toothpicks.

Similarly, each column has one more horizontal toothpick than there are squares. Thus a column that has 6 squares down has 7 horizontal toothpicks. Since the rectangle has 8 columns, there is a total of $8 \times 7 = 56$ horizontal toothpicks. Thus, $54 + 56 = $ **110 toothpicks are needed to form the 6-square by 8-square rectangle.**

METHOD 2:   *Strategy: Group toothpick figures conveniently.*
Many other ways of grouping toothpicks are possible. For example, starting from the upper left corner, there are $6 \times 8 = 48$ figures of the shape $\lceil$. That requires 96 toothpicks. There are $6 + 8 = 14$ more toothpicks along the right side and bottom of the rectangle, for a total of $96 + 14 = 110$ toothpicks.

Or, the top row of squares contains 25 toothpicks, and each of the remaining five rows contains 17 toothpicks. Then $25 + 5 \times 17 = 25 + 85 = 110$ toothpicks.

FOLLOW-UPS: *(1) How many different methods can your students find to solve this problem? (2) Develop a formula to find the number of toothpicks needed to form an N-square by N-square rectangle. [$2N^2 + 2N$; various methods yield other equivalent expressions.]*

SET 13   **Olympiad 5**

5A.   *Strategy: Match percent full to amount held.*
A tank that is 80% empty must be 20% full. Since 20% of the tank holds 11 gallons, then 100% of the tank holds 5 times as much. Thus, **the full tank holds** $5 \times 11 =$ **55 gallons.**

FOLLOW-UP: *If a tank contains 25 gallons when it is 90% empty, how many gallons does it contain when it is 30% empty? [175]*

5B.   *Strategy: Follow the rules for smaller numbers and look for a pattern.*
The table contains Ashley's first 12 numbers. Notice that the numbers 24, 8, 32 and 6 in that order repeat themselves endlessly. Thus, the 4th, 8th, 12th and every fourth number after that has a value of 6. Since 100 is a multiple of 4, **the 100th number to be inserted is 6.**

1.	**1**	given	6.	**8**	by Rule 2	11.	**32**	by Rule 1
2.	**4**	by Rule 1	7.	**32**	by Rule 1	12.	**6**	by Rule 2
3.	**16**	by Rule 1	8.	**6**	by Rule 2			
4.	**6**	by Rule 2	9.	**24**	by Rule 1			
5.	**24**	by Rule 1	10.	**8**	by Rule 2			

5C.   *Strategy: Relate the radii of the small circles to that of the large circle.*
Since the area of each small circle is 4π, the radius of each small circle is 2. Then the diameter of the large circle is 12 and its radius is 6. Thus the area of the large circle is 36π, and since $N\pi$ represents 36π, N **represents 36.** *(Note: The answer of 36π did not receive credit since its value is about 113.10, not 36.)*

FOLLOW-UP: *A rectangle is inscribed in a quadrant of a circle. If the length of the diagonal shown is 7, what is the area of the circle? [49π]*

5D. METHOD 1: *Strategy: Compare the differences in their daily amounts.*
On Day 1, Josh saves 29 cents less than Sarah.
On Day 2, Josh saves 28 cents less than Sarah.
On Day 3, Josh saves 27 cents less than Sarah.
Therefore, for the first 29 days, Josh saves $29 + 28 + 27 + ... + 1$ cents less than Sarah.
Then, on Day 30, both people save the same amount.
Finally, during the next 29 days, Josh saves $1 + 2 + 3 + ... + 29$ cents *more* than Sarah. Therefore, the deficit Josh accumulates during the first 29 days equals the surplus he accumulates during the last 29 days. **At the end of the** $29 + 1 + 29 = 59$ **days they will have saved the same total amounts.**

METHOD 2: *Strategy: Use Gauss' method, and algebra.*
Sarah saves 30¢ a day for n days, for a total of $30n$ cents.
Josh saves $1 + 2 + 3 + ... + (n-2) + (n-1) + (n)$ cents, in n days.
Add them together in pairs, as shown:

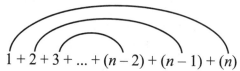

$$1 + 2 + 3 + ... + (n-2) + (n-1) + (n)$$

There are $\frac{n}{2}$ pairs, each with a sum of $n + 1$, for a total of $\frac{n}{2}(n + 1)$ cents.
Thus Josh saves $\frac{n}{2}(n + 1)$ cents and Sarah saves $30n$ cents.
After n days, Sarah and Josh have saved the same amount in cents: $\frac{n}{2}(n + 1) = 30n$

$$\frac{n}{2}(n + 1) = 30n \quad \text{[Double both sides of the equation.]}$$
$$n(n + 1) = 60n \quad \text{[Divide both sides of the equation by the number } n.]$$
$$n + 1 = 60 \quad \text{[Subtract 1 from both sides of the equation.]}$$
$$n = 59 \quad \text{At the end of Day 59, they have saved the same amount.}$$

FOLLOW-UP: How is this problem similar to 4C?

5E. *Strategy: Compute the least common multiple (LCM) of the three numbers.*
The Least Common Multiple (LCM) of 3, 5, and 7 is 105. Since an odd number is never divisible by an even number, 105 is not divisible by 4, 6, or 8. Then **the least such number is 105.**

FOLLOW-UPS: (1) What are the next 4 numbers that satisfy the given conditions? [315, 525, 735, and 945] *(2) If the word "positive" is omitted from the question, how is the problem changed?*

SOLUTIONS: DIVISION M

1A. *Strategy:* *Evaluate each term separately.*
The value of 1 raised to any integral power is still 1. For example, $1^4 = 1 \times 1 \times 1 \times 1 = 1$. Since there are only even powers of 1, there are 50 addends of 1 to be summed. Therefore **the sum is 50 × 1 = 50**.

FOLLOW-UP: *What is the value of $(-1)^1 + (-1)^2 + (-1)^3 + \ldots + (-1)^{100} = ?$ [0]*

1B. **METHOD 1:** *Strategy: Examine the common multiples of 3 and 4.*
The least common denominator of $\frac{1}{3}$ and $\frac{1}{4}$ is 12. Only multiples of 12 will be divisible by both 3 and 4. The actual number of whole apples therefore is a multiple of 12.

Suppose for the moment that the total number of apples picked is 12. Then Lenny would pick 4 apples, Jenny would pick 3 apples, and the difference would be 1. However, it is given that the actual difference is 7, not 1. Therefore, multiply all quantities by 7: Lenny picks 28 apples, Jenny picks 21 apples, for a total of 49 apples. The trio picks a total of $7 \times 12 = 84$ apples and **Penny picks** $84 - 49 =$ **35 apples**.

METHOD 2: *Strategy: Use algebra.*
Let $x =$ the total number of apples picked.
Then Jenny picks $\frac{1}{3}x$ apples and Lenny picks $\frac{1}{4}x$ apples.

$$\frac{1}{3}x - \frac{1}{4}x = 7$$
$$\frac{4}{12}x - \frac{3}{12}x = 7$$
$$\frac{1}{12}x = 7$$

Multiply each side of the equation by 12 to get $x = 84$.

Therefore, the total number of apples picked is 84, Jenny picks $\frac{1}{3}$ of $84 = 28$ apples, Lenny picks $\frac{1}{4}$ of $84 = 21$ apples, and Penny picks $84 - (28 + 21) = 35$ apples.

METHOD 3: *Strategy: Consider each person's fraction of the total.*
Jenny and Lenny together pick $\frac{1}{3} + \frac{1}{4} = \frac{7}{12}$ of the apples. Thus Penny picks the remaining $\frac{5}{12}$ of the apples. Lenny picks $\frac{1}{3} - \frac{1}{4} = \frac{1}{12}$ more of the apples than Jenny, which represents 7 apples. So the total number of apples picked is $12 \times 7 = 84$. Then Penny picks $\frac{5}{12}$ of $84 = 35$ apples.

1C. **METHOD 1:** *Strategy: Track the dollar changes.*

Month	Before Change	Change (%)	Change ($)	After Change
October	$1,000	20%	$200	$1,200
November	$1,200	−20%	−$240	$960

The November price is $960.

METHOD 2: *Strategy: Track the percent changes.*
Note that in Method 1, $1200 is 120% of the $1000. Increasing an amount by 20% is equivalent to finding 120% of that amount, so that the October price is 120% of the September price. Then $1.20 × $1,000 = $1,200. Similarly, a decrease of 20% is equivalent to finding 80% of the amount. Then 80% of $1200 = $960.

METHOD 3: *Strategy: Replace the two percents with a single percent.*
The October price is 120% of the September price, as shown in Method 2, and the November price is 80% of the October price. Then the November price is 80% of 120% = 96% of the September price: 96% of $1000 is $960.

FOLLOW-UPS: (1) What is the November price if the order of the percent changes is reversed? [also $960] (2) What single percent reduction can replace three consecutive discounts of 10%, 20% and 25%. [46%: one way is to use .90 × .80 × .75 = 54% of the original price.]

1D. *Strategy: Consider the fewest number of correct answers: check even vs. odd.*
To score 59 points Jana needs to get at least 12 questions right (5 × 12 = 60). Because 59 is odd but the total deduction (at 2 points each) is even, the number of correct answers (at 5 points each) is odd. Thus Jana has 13, 15, 17, or 19 correct answers. Suppose Jana has 13 correct answers: 5 × 13 = 65. Then the deduction would be 65 – 59 = 6 points for 3 incorrect answers. Therefore **Jana omitted** 20 – 13 – 3 = **4 questions.**

Do any other answers produce a score of 59? For every two additional correct answers (10 additional points), Jana needs 5 additional incorrect answers (10 points less) to maintain the score of 59. Then these seven additional questions produce a total of at least twenty-three questions. Since only 20 questions are asked, just one way exists to score 59 points.

FOLLOW-UPS: (1) Of 31, 48, 89, and 100, which point total would be impossible on the above game? [89] (2) What other positive point totals are impossible to get? [77, 82, 84, 87, 89, 91, 92, 94, 96-99] (3) How many different ways are there to score exactly 30 points? [3]

1E. *Strategy: Use the fact that the hour hand moves proportionately to the minute hand.*
In any one hour the hour hand sweeps through 360 ÷ 12 = 30 degrees. If, at 2:20, the hour hand were pointing directly at the 2 and the minute hand at the 4, the angle formed by the hands would be 60 degrees. However at 2:20, since the minute hand has moved through $\frac{1}{3}$ of a rotation, the hour hand sweeps through $\frac{1}{3}$ of the 30° between 2 and 3 o'clock, that is through 10°. **The angle between the hands of a clock at 2:20 is** 60 – 10 = **50°.** Alternately, the same result can be achieved by adding $\frac{2}{3}$ of the 30° between 2 and 3 o'clock to the 30° between 3 and 4 o'clock. Or, consider the 12 noon as a 0° position. Then the minute hand is at the 120° position and the hour hand is at the 30 + 30 + 10 = 70° position. The difference in their positions is 50°.

FOLLOW-UP: How many times will the hour and minute hands cross each other (coincide) in a 12 hour period? [Be careful - the only time they cross each other between the hours of 11:00 and 1:00 is at exactly 12:00. The answer is 11 times]

2A. *Strategy: Pair the multiples so as to obtain a constant sum.*

The set of two digit multiples of 4 is {12, 16, 20, ... 52, 56, ... 88, 92, 96}.

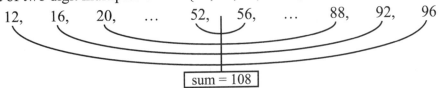

Notice that if we pair the numbers as shown above, the sum of each pair of numbers remains constant, namely 108. This is because as we switch from one pair to the next, the first member is increased by 4 and the second member is decreased by 4. Thus, the average of the entire set is the average of any one of these pairs, namely 54. **The arithmetic mean of all the positive two-digit multiples of 4 is 54.**

FOLLOW-UP: What is the average of all positive and negative 2-digit multiples of 4? [0]

2B. *Strategy: Examine the digits from left to right.*

The form of the palindrome is *ABCBA*. Since 0 is never a leading digit, *A* (on the left) can be any digit from 1 through 9, and *B* (on the left) can be any digit from 0 through 9. For each of the 9 possible values of *A*, there are 10 possible values of *B*, a total of $9 \times 10 = 90$ different 2-digit numbers *AB*. Similarly, *C* can be any digit from 0 through 9. For each of the 90 possible values of *AB*, there are 10 possible values of *C*, a total of $90 \times 10 = 900$ different values of the 3-digit *ABC*. Since *B* and *A* (on the right) each have only 1 possible value, the same as the one on the left, **there is a total of** $9 \times 10 \times 10 \times 1 \times 1 = $ **900 5-digit palindromes.**

FOLLOW-UPS: (1) How many palindromes exist that have 1 digit? 2? 3? 4? 5? 6? [10 (including 0), 9, 90, 90, 900, 900] (2) How many palindromes exist between 1 million and 1 billion? [108,000]

2C. **METHOD 1:** *Strategy: Reason, based on the differences of $11 and $5.*

The total difference between $11 over the cost and $5 under the cost is $16, and it is caused by a per-share difference of $8 – $6 = $2 per person. Thus there are $16 \div 2 = 8$ friends. If 8 friends pay $8 each and the total is over by $11, **the price of the video game is $53.** To check, multiply $6 by 8 people and then add $5.

METHOD 2: *Strategy: Construct a table comparing cost in each of the two ways.*

In the table, row 2 shows the game's cost for 1 or 2 or 3 or ... friends, if each one pays $8. Row 3 shows the cost if each pays $6. Check for the same entry in both rows.

Cost of Video Game:

Number of friends	1	2	3	4	5	6	7	8
If each chips in $8	–	5	13	21	29	37	45	53
If each chips in $6	11	17	23	29	35	41	47	53

Continued.

The price of the video game is $53. The conditions of the problem hold for 8 friends.

FOLLOW-UP: The cost of 3 equally priced shirts and 2 equally priced ties is $103. The cost of 2 of these shirts and 3 of these ties is $97. What is the cost in dollars of one of the shirts? [23]

2D. _Strategy: Use the distributive property._

Notice that $\frac{7}{19}$ is a factor of each of the four terms. Then $\frac{7}{19} \times \frac{13}{44} + \frac{7}{19} \times \frac{19}{44} + \frac{7}{19} \times \frac{25}{44} + \frac{7}{19} \times \frac{31}{44}$ can be rewritten as $\frac{7}{19} \times (\frac{13}{44} + \frac{19}{44} + \frac{25}{44} + \frac{31}{44})$. The sum of the fractions in the parentheses is $\frac{88}{44}$, or 2. Then **the value of N is** $\frac{7}{19} \times 2 = \frac{14}{19}$.

(Note 1: You may also use the distributive property more comprehensively:

$(\frac{7}{19} \times \frac{1}{44}) (13 + 19 + 25 + 31) = \frac{7}{19} \times \frac{1}{44} \times 88 = \frac{14}{19}$

(Note 2: MOEMS never assigns a problem that can only be done with huge amounts of computation. With problems like 2D, students should be encouraged to look for another approach, one that takes advantage of the way the numbers are arranged.)

FOLLOW-UP: Use the distributive property to simplify the product of $1365 \times 468 - 234 \times 2720$. *[2340]*

2E. The length of one side of each tile is $\frac{3}{4}$ of a foot. Thus, the floor is 18 feet $\div \frac{3}{4}$ foot = 24 tiles long and $15 \div \frac{3}{4}$ = 20 tiles wide.

METHOD 1: _Strategy: Convert to tiles and then subtract areas._
See *figure 1*. The entire floor consists of $24 \times 20 = 480$ tiles. The blue portion of the floor consists of $22 \times 18 = 396$ tiles. Thus, $480 - 396 =$ **84 tiles are white.**

Figure 1

METHOD 2A: _Strategy: Count tiles along each edge._
See *figure 2a*. Counting tiles along the border produces $24 + 20 + 24 + 20 = 88$ tiles. However, each of the 4 corner tiles was counted twice. Thus, 84 tiles are white.

Figure 2a

METHOD 2B: _Strategy: Consider the longer and shorter sides separately._
See *figure 2b*. The shorter two sides of the border contain 20 tiles each. This includes the corner tiles. Thus, the longer two sides contain $24 - 2 = 22$ uncounted tiles each. Then the border contains $20 + 20 + 22 + 22 = 84$ white tiles.

Figure 2b

FOLLOW-UPS: (1) Suppose a second border of white tiles is placed around the existing border. What percent of the tiles now are white? [30.77%] (2) Suppose the 480 tiles alternated in color: red, then white, then blue, repeating continuously from left to right and then from top to bottom. If the tile in the upper left corner is red, what color is the tile in the lower right corner? [blue]

SOLUTIONS: DIVISION M

3A. _Strategy: Use the Commutative Property, then separate the fractions._
Rewriting the denominator (or the numerator) we have:

$$\frac{28 \times 26 \times 24 \times 22}{11 \times 12 \times 13 \times 14} = \frac{22 \times 24 \times 26 \times 28}{11 \times 12 \times 13 \times 14} = \frac{22}{11} \times \frac{24}{12} \times \frac{26}{13} \times \frac{28}{14} = 2 \times 2 \times 2 \times 2 = 2^4 = \mathbf{16}.$$

FOLLOW-UPS: _(1) If $30 \times 40 \times 50 \times 60 \times 70 = 3 \times 4 \times 5 \times 6 \times 7 \times N$, what is the value of the number N?_
[100,000] (2) Simplify a fraction whose denominator is $1 \times 2 \times 3 \times 4 \times 5$ and whose numerator is
$25 \times 16 \times 9 \times 4 \times 1$. [120] (3) Use these ideas to introduce factorials and perfect squares.

3B. _Strategy: Examine groups of pages separately._
Number of pages written using the digit "1":

Page Number	Number of pages
1 - 9	1
10 - 19	10
20 - 99	8
100 - 177	78
Total	**97**

If all outcomes are equally likely, a probability is a fraction in which the denominator is the
number of total outcomes and the numerator is the number of favorable outcomes. In this case
there are 177 equally likely total outcomes (the total number of pages) and there are 97 favorable
outcomes (the number of pages whose page number contains the digit "1" at least once), as
shown in the table. **The probability that a page number contains the digit "1" is $\frac{97}{177}$.**

FOLLOW-UPS: _(1) How many times in all does the digit "1" occur among the same page numbers? [116,_
counting by place value] (2) Which digits appear the fewest number of times? [0,9,8] (3) How many
times does each appear? [27]

3C. _Strategy: Draw a diagram and examine the intervals._
The distance between streetlights 1 and 3 is 600
meters. Since there are two intervals between
these lights, the distance in each interval is 300
m. Since 15 streetlights have fourteen intervals,
**the distance between the first and fifteenth
streetlights is** $14 \times 300 = \mathbf{4200}$ **m.**

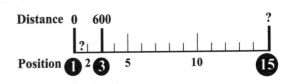

FOLLOW-UP: _A carpenter saws a board into 5 equal pieces in 10 minutes. At the same rate, how long_
does it take to saw a board into 10 equal pieces? [22.5 minutes]

SOLUTIONS: DIVISION M

3D. **METHOD 1:** *Strategy: Make a table of simpler cases and examine the totals.*
Draw a triangle with 1 row. Count the total number of unit triangles. Repeat for 2, 3, and 4 rows.

Row Number	1	2	3	4	...	12
Number of Triangles in Row	1	3	5	7	...	?
Total Number of Triangles	1	4	9	16	...	?

The third line of the table totals all the numbers in the second line to that point.

Every entry in the third line of the table is a perfect square. The first row contains $1 = 1^2$ triangles, the first two rows contain $4 = 2^2$ triangles, the first three rows contain $9 = 3^2$ triangles, and so on. Extending the pattern, **the first 12 rows contain $12^2 = $ 144 unit triangles.**

METHOD 2: *Strategy: Make a table of simpler cases and examine each row separately.*
All the entries in the second line of the table above are consecutive odd numbers. The first row has 1 triangle, the second has 3 triangles, the third has 5, and so on. Each additional row has two more triangles than the row before. Thus we need to add the first twelve odd numbers: $1 + 3 + 5 + 7 + ... + 19 + 21 + 23$. One way to add easily is to group them: $(1 + 23) + (3 + 21) + (5 + 19) + ... + (11 + 13) = 6 \times 24 = 144$.

FOLLOW-UPS: *(1) How many unit triangles are in rows 11 - 20? In rows 31 - 40? In row k? [300; $1600 - 900 = 700$; $2k - 1$] (2) What is the sum of the first 700 consecutive odd numbers? [490,000] (3) How many triangles of all sizes are contained in the first four rows? [26]*

3E. *Strategy: Use the fact that the average rate = total distance ÷ total time.*
In the table below, multiply left to right and add down. Thus, each distance is the product of the rate and time, and the total time and distance each is the sum of the individual legs of the trip. Rates cannot be added.

	Rate	Time	Distance
Going	40	?	120
Returning	M	?	?
Whole trip	48	?	?

The total trip is 240 miles, 120 miles each way. Since the average for the whole trip is 48 mph, the time for the whole trip is $240 ÷ 48 = 5$ hours.

The trip from Antwerp to Brussels takes 120 miles ÷ 40 mph = 3 hours. So the return trip of 120 miles takes $5 - 3 = 2$ hours. Hence **the value of M is 60 mph.** The table shows this information efficiently.

Rate	Time	Distance
40	3	120
M	2	120
48	5	240

FOLLOW-UPS: *(1) What would the rate of the return trip be if the given distance were 200 miles? 500 miles? 960 miles? [60 mph in each case] (2) The average of all 10 of Janine's tests is 90. The average of Janine's first 6 tests is 88. What is the average of her last 4 tests? [93]*

SOLUTIONS: DIVISION M

4A. _Strategy: Consider one condition at a time._
Ten two-digit choices exist in which the sum of the tens and units digits is 9. Since the number is even, choices 09, 27, 45, 63, and 81 are eliminated, and five choices for the last two digits remain: 90, 72, 54, 36 or 18.

Because the number is divisible by 3, the sum of the digits is a multiple of 3; hence 12, 15, or 18. Therefore, there are three choices for the hundreds digit; 3, 6, or 9. Thus, $5 \times 3 = $ **15 three-digit numbers satisfy all of the conditions.**

(**Note:** Some of the above may by replaced by making an organized list of the choices.)

4B. _Strategy: Substitute and simplify._
First evaluate $3 \succ 5$: $3 \succ 5 = \frac{3 + 5}{3 - 5} = \frac{8}{-2} = -4$

Then evaluate $5 \succ 3$: $5 \succ 3 = \frac{5 + 3}{5 - 3} = \frac{8}{2} = +4$

Therefore, $\frac{3 \succ 5}{5 \succ 3} = \frac{-4}{4} = -1$.

The value is –1.

**Follow-Ups:** (1) Without a calculator, compute the value of $\frac{164 \succ 468}{468 \succ 164}$. [–1]
(2) Compute the value of $(\frac{5}{4} \succ \frac{5}{6}) \succ \frac{5}{8}$. [$\frac{9}{7}$]

4C. **METHOD 1:**   _Strategy: Use the radii to associate the side-lengths of the two squares._
Since the area of the outer square is 100 square cm, the length of one side of that square is 10 cm. Therefore, the diameter of each of the congruent circles measures 5 cm, and the radius 2.5 cm. The inner square is composed of 8 connecting radii; two of them form a side of the inner square. The length of this side is 5 cm. Then **the area of the inner square is 25 sq cm.**

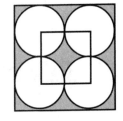

METHOD 2:   _Strategy: Act it out using coins._
Think of the large square as a 4-penny square. The small square is then 4 fourths of a penny, or a 1 penny square. Therefore, the area of the small square is $\frac{1}{4}$ of that of the large square, and is equal to $100 \div 4 = 25$ sq cm.

**Follow-Ups:** (1) What is the combined area of the shaded regions? [$100 - 25\pi$] (2) Using $\pi = 3.1416$, what percent of the outer square is shaded? [21.46%] (3) What percent of the outer square would be shaded if there were 1, 9, 16, or 25 congruent circles? [Also 21.46%]

4D. <u>*Strategy*</u>: *Compute each region separately using a Venn Diagram.*

For each region, use the letter V to represent those who like volleyball, S for those who like softball, and B for those who like basketball. (These letters are not variables, but labels to identify regions.)

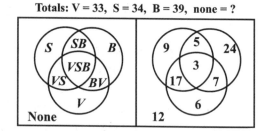

Totals: V = 33, S = 34, B = 39, none = ?

First assign the value 3 to the region VSB which identifies those who like all three sports. The region VSB is common to all three circles.

Then assign values to those regions that identify those who like 2 sports but not the third. Place $20 - 3 = 17$ in the region VS, which identifies the set of those who like volleyball and softball but not basketball. Similarly, place $10 - 3 = 7$ into region BV and $8 - 3 = 5$ into region SB.

Next, assign values to those regions that identify those who like one and only one sport. Into region V place $33 - 17 - 7 - 3 = 6$ to identify those who like volleyball but neither of the other sports. Similarly, into region S place $34 - 17 - 5 - 3 = 9$ and into region B place $39 - 7 - 5 - 3 = 24$. Finally, put together all the pieces, remembering to include the 12 people who like none of the sports: $(3) + (17 + 7 + 5) + (6 + 9 + 24) + (12) =$ **83 students are in the 8th grade**.

4E. <u>*Strategy*</u>: *Work with the denominators.*

The denominators of the two fractions are factors of 18. Factor pairs include 1 and 18, 2 and 9, 3 and 6, and also such as 6 and 9, 2 and 18, 3 and 18, and so on. Eliminate 1 and 18 because the value of $\frac{18}{1} - \frac{1}{18}$ will be much larger than $\frac{77}{18}$. Eliminate 3 and 6 (and also such as 6 and 9, etc.) because $\frac{3}{6}$ (also $\frac{6}{9}$) is not in lowest terms. Therefore only the remaining pair 2 and 9 yields the desired difference. **The proper fraction is $\frac{2}{9}$.** Check: $\frac{9}{2} - \frac{2}{9} = \frac{81}{18} - \frac{4}{18} = \frac{77}{18}$.

FOLLOW-UP: *If a proper fraction in lowest terms is subtracted from its reciprocal, the difference is $\frac{299}{90}$. What is the proper fraction?* $[\frac{5}{18}]$

5A. <u>*Strategy*</u>: *Examine the ones place.*

Note that in the ones' place, $A \times A$ ends in A. This is true only for the digits 0, 1, 5, or 6. But if $A = 0$ or 1, *TADA* is not possible. However, if $A = 5$, then *TADA* represents 25_5 so that $H = 1$ (to avoid "carrying" into the hundreds place) and $D = 7$. On the other hand, if $A = 6$, then *TADA* = 36_6 so $H = 1$ again and $D = 9$. Either way, **H represents the digit 1**.

FOLLOW-UP: *In the division at the right, what is the quotient?* [13]

```
        ? ?
  2 ? ) 3 ? 8
        2 ? ?
        ─────
          7 8
          ? ?
          ───
            0
```

5B. _Strategy_: Use the Order of Operations: Multiply first, then add.

Term	Value
(-2)	-2
$(-2)(-2)$	$+4$
$(-2)(-2)(-2)$	-8
$(-2)(-2)(-2)(-2)$	$+16$
$(-2)(-2)(-2)(-2)(-2)$	$\underline{-32}$
Total	**-22**

To add, group by sign: $({}^+4 + {}^+16) + ({}^-2 + {}^-8 + {}^-32) = {}^+20 + {}^-42 = {}^-22$.
The value of the given expression is -22.

FOLLOW-UPS: _(1) What is the value of $(^-3)^2 - (^-3)^3 - (^-3)^4 - (^-3)^5$? [+198] (2) Then: what is the value of $(^-10)(^-3)^2 - (^-10)(^-3)^3 - (^-10)(^-3)^4 - (^-10)(^-3)^5$? [-1980, by distributive property] (3) Is the value of $(^-2) + (^-2)^2 + (^-2)^3 + ... + (^-2)^{50}$ positive or negative? [Positive; each even power has a greater absolute value than the preceding odd power.]_

5C.   **METHOD 1:**   _Strategy_: Use the Transitivity Principle.

Note that 8 bears weigh as much as 15 cougars. Since 4 bears weigh as much as 9 apes, then 8 bears also weigh as much as 18 apes. Therefore, 15 cougars weigh as much as 18 apes, which means that 5 cougars weigh as much as 6 apes.

Note that 10 cougars weigh as much as 27 deer. But 10 cougars also weigh as much as 12 apes, so 27 deer weigh as much as 12 apes. **Simplifying, 9 deer weigh the same as 4 apes.** The following table displays this information efficiently:

	A	B	C	D	
Start	9	4			$\underline{?}A = 8B$
Step 1	18	8	15		$\underline{?}A = 10C$
Step 2	12		10	27	$4A = \underline{?}D$
Step 3	4			(9)	Answer = 9

METHOD 2:   _Strategy_: Use algebra.

Suppose the weight of one ape, bear, cougar, and deer are represented by a, b, c, and d, respectively. Then the equations are $9a = 4b$, $8b = 15c$, and $10c = 27d$. Rewrite the first equation as $18a = 8b$. Since both $15c$ and $18a$ are equal to $8b$, then $15c = 18a$, which simplifies to $5c = 6a$. Since $10c = 27d$, rewrite $5c = 6a$ as $10c = 12a$. Since $10c$ is equal to both $27d$ and $12a$, then $27d = 12a$, which simplifies to $9d = 4a$. Thus, 9 deer weigh the same as 4 apes.

(Note: The thinking in Methods 1 and 2 are identical.)

Method 3 is on next page.

SOLUTIONS: DIVISION M

METHOD 3:   _Strategy: Assume a convenient weight since all relationships are ratios._
Suppose each ape weighs, say, 20 kg (With ratios, the number chosen need not be accurate but should allow ease of computation). Then the question becomes: How many deer weigh 80 kg? The chain of calculations would be as follows:

1. If 9 apes weigh a total of 180 kg, then 4 bears also weigh 180 kg. Thus, 1 bear weighs 45 kg.
2. Since 8 bears weigh a total of 360 kg, then 15 cougars also weigh 360 kg. Thus 1 cougar weighs 24 kg.
3. Since 10 cougars weigh a total of 240 kg, then 27 deer also weigh 240 kg. Thus 1 deer weighs $\frac{240}{27} = \frac{80}{9}$ kg.
4. Thus, it takes 9 deer to weigh 80 kg, so that 4 apes would weigh the same as 9 deer.

5D. _Strategy: First find the area of triangle OED._
It can be shown that if side $\overline{ED}$ is considered the base of triangle AED, then side $\overline{EA}$ will be the altitude to that base.

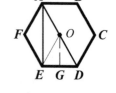

In triangle OED, draw the altitude $\overline{OG}$ to base $\overline{ED}$. The length of $\overline{OG}$ is half the length of $\overline{EA}$. Note that triangles AED and OED share the same base $\overline{ED}$, but the altitude of triangle OED is half the altitude of triangle AED. Since the area of a triangle is one-half the product of its altitude and base, the area of triangle OED is one-half the area of triangle AED. The area of triangle AED is given as 12 sq cm, so the area of triangle OED is 6 sq cm.

A regular hexagon can be broken up into 6 congruent equilateral triangles, as shown in figure 2. Since the area of one of the triangles, OED, is 6 sq cm, **the area of the hexagon is 36 sq cm.**

Figure 2

FOLLOW-UPS: _(1) What is the area of triangle AOE? Triangle AFE? Trapezoid ABCD?_ [6, 6, 18] _(2) Suppose the sides of a right triangle are 3 cm, 4 cm, and 5 cm, and the sides of a similar right triangle are 33 cm, 44 cm, and 55 cm. How many times the area of the first triangle is the area of the second triangle?_ [121]

5E. _Strategy: Use the fact that the factors of 10 are 5 and 2._
Noting that 2^{20} can be written as:

$(2 \times 2 \times 2 \times 2 \times 2 \times 2 \times 2 \times 2 \times 2 \times 2 \times 2 \times 2 \times 2 \times 2 \times 2 \times 2 \times 2 \times 2) \times (2 \times 2)$, then $5^{18} \times 2^{20} = 5^{18} \times 2^{18} \times 2^2$. Since $5^{18} \times 2^{18}$ is 1 followed by a string of 18 zeroes, then **the leading digit is** $1 \times 2^2 = \mathbf{4.}$

FOLLOW-UPS: _(1) What is the leading digit in the product of $25^9 \times 4^{10}$?_ [4; this is the same problem as 5E] _(2) What are the two leading digits in the product of $5^{20} \times 2^{18}$?_ [25] _(3) How many terminal zeroes does the product in 30! contain?_ [7]

SOLUTIONS: DIVISION M

1A. **METHOD 1:**   *Strategy: Count up from 2001 to produce a multiple of 11.*
If 2001 is divided by 11, the remainder is <u>10</u> (and the quotient is 181). If <u>1</u> is added to 2001, 2002 will be a multiple of 11. **The first year in the 21st century that is divisible by 11 is 2002.**

Number	Sum of digits in odd numbered places	Sum of digits in even numbered places	Difference	Multiple of 11?
2001	2 + 0 = 2	0 + 1 = 1	1	No
2002	2 + 0 = 2	0 + 2 = 2	0	Yes (0×11)

METHOD 2:   *Strategy: Use a test for divisibility by 11.*
Find the sum of the digits in the odd numbered places, and then the sum of the digits in the even numbered positions. The original number is a multiple of 11 if and only if the difference in these sums is a multiple of 11. The first year in the 21st century that is divisible by 11 is 2002.

METHOD 3:   *Strategy: Do the division.*
Exactly one of any 11 consecutive integers is divisible by 11. Try dividing 2001, 2002, 2003, …, 2011 in turn by 11, starting with 2001. 2002 is divisible by 11, so the first year in the 21st century that is divisible by 11 is 2002.

FOLLOW-UPS: (1) The four-digit whole number 7A98 is divisible by 23. Find the missing digit A. [4]
(2) The five-digit whole number 4B5B7 is divisible by 11. Find the value of B. [8]

1B. **METHOD 1**   *Strategy: Work from the outside in.*
The next palindrome after 20902 is greater than 21000. If the first two digits are 21, the last two digits are 12. The least middle digit is 0. **The next palindrome is 21012.**

METHOD 2:   *Strategy: Work from the inside out.*
Consider the number formed by the middle 3 digits, 090. The next greater 3-digit palindrome is 101. The outer digits need not change. The next greater 5-digit palindrome is 21012.

1C. **METHOD 1:**   *Strategy: Count triangles horizontally in an organized way.*
Every triangle in the figure has one vertex at *A*. Each horizontal segment serves as the base of a triangle. Segment $\overline{DE}$ contains 3 segments of "length" 1, 2 segments of "length" 2, and one segment of "length" 3, giving a total of 6 different triangles. Similarly, $\overline{FG}$ and $\overline{BC}$ each contain the bases of 6 different triangles. **The total number of triangles in the figure is $3 \times 6 = 18$.**

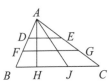

METHOD 2: *Strategy: Count triangles vertically.*

The triangular region *ABH* contains 3 different triangles of "width" 1. The same is true for triangular regions *AHJ* and *AJC*. A total of 9 triangles are therefore 1 "unit wide". The triangular region *ABJ* contains 3 different triangles of "width" 2. The same is true for triangular region *AHC*. A total of 6 triangles are therefore 2 "units wide". The triangular region *ABC* contains 3 different triangles of "width" 3. A total of 3 triangles are therefore 3 "units wide". The total number of triangles in the figure is 9 + 6 + 3, or 18.

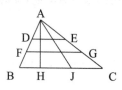

FOLLOW-UP: Suppose 1, 2, 3, ... , 20 additional horizontal line segments are drawn within triangle ABC, making 4, 5, 6, ..., 23 horizontal line segments in all. How many triangles of any size are there? [24, 30, 36, ..., 138]

1D. **METHOD 1:** *Strategy: Consider the deviations from the average.*

Suppose the only students in the class are Amanda, Barb, and Colin. The sum of Amanda's and Barb's scores is 3 points more than the sum would be if each earned the average score. Colin's score, 82, is therefore 3 points less than the average score. **The class average is 85.**

METHOD 2: *Strategy: Consider the points scored.*

Let □ be the class average.

$$(\square - 5) + (\square + 8) + 82 = \square + \square + \square$$
$$\boxtimes + \boxtimes + 85 = \boxtimes + \boxtimes + \square$$
$$85 = \square \qquad \text{The class average is 85.}$$

METHOD 3: *Strategy: Use algebra (for more advanced students).*

Let *A* be the class average.

Then *A* − 5 = Amanda's score and *A* + 8 = Barb's score.

$$\frac{(A-5)+(A+8)+82}{3} = A$$

Solving, *A* = 85. The class average is 85.

1E.

Strategy: Organize according to the number of digits.

The first 9 pages require nine digits. Pages 10-99 contain two digits each and there are 90 of these pages, hence 180 additional digits. There are 228 − 9 − 180 = 39 digits left to number the pages following page 99. Every page from 100 through 999 inclusive requires three digits, so the diary has 39 ÷ 3 = 13 more pages. Then, 99 + 13 = **112 pages are in Joshua's diary.**

SOLUTIONS: DIVISION M

2A. *Strategy: Examine the hundreds column first.*
Since the sum of any two 2-digit numbers is less than 200, *B* is 0 or 1. Because the leading digit is never 0, *B* is 1. In the units column, *A* + 4 ends in a 1. **The value of *A* is 7.**

2B. *Strategy: Use the fact that 2 is the only even prime.*
2001 is an odd number. If the sum of two numbers is odd, one addend is odd and the other is even. The only even prime is 2, so the two addends are 2 and 1999. **The greater prime is 1999.**

FOLLOW-UPS: (1) To verify that 1999 is prime, it must be shown that 1999 is not a multiple of any prime from 2 through N. What is the least possible value of N? [43, the greatest prime whose square is not greater than 1999] (2) The sum of two prime numbers is 102. Find all such pairs of primes. [5 & 97; 13 & 89; 19 & 83; 23 & 79; 29 & 73; 31 & 71; 41 & 61; 43 & 59]

2C. *Strategy: Compare the two series term by term.*
Increasing every term in the first series by 30 results in the second series. Each series has 30 terms. Therefore, increasing the given sum, 465, by 30 × 30 = 900 yields the desired sum. ***N* is 1365.**
(Note: This can also be thought of as translating the entire series upwards by 30)

2D. *Strategy: Determine the days on which each could make the statement.*
Suppose a "truthteller" says "Tomorrow I will lie." The statement is true. Suppose a liar says "Tomorrow I will lie." This statement is false, so tomorrow the speaker will tell the truth. *In either case, the speaker's "truth status" is different today from what it will be tomorrow.*
The chart below gives the "truth status" for each speaker on each day of the week.

Speaker	Sun	Mon	Tues	Wed	Thurs	Fri	Sat
Maxie	F	T	T	T	T	F	F
Minnie	T	T	F	F	F	T	T

The only two consecutive days in which the truth status changes for *both* speakers is from Thursday to Friday. **The day on which both could say "Tomorrow I will lie" is Thursday.**

SOLUTIONS: DIVISION M

2E. *Strategy: Split the figure into congruent triangles.*
Draw $\overline{GH}$ as shown. The entire figure consists of six congruent triangles, two
of which are shaded. The shaded region has an area of 12 cm², so each of the
six congruent triangles has an area of 6 cm². Triangle *ABC* is composed of four
of the smaller triangles, and **the area of triangle *ABC* is 24 cm².**

FOLLOW-UPS: *(1) Suppose in the figure $\overline{FB}$ is drawn. What is the area of triangle
FGB?* [6 cm²] *(2) What is the ratio of the areas of parallelogram AEDB to trapezoid GHCB?* [4:3]
(3) How does making the two original triangles scalene affect the solution? [no change]

SET 15	Olympiad 3

3A. *Strategy: Find 2 consecutive numbers whose product ends in zero.*
The page numbers differ by 1. The product ends in 0, so the units digits of the page numbers will
be either 0 and 1, 4 and 5, 5 and 6, or 9 and 0. They must be close to 20 because $20 \times 20 = 400$.
$20 \times 21 = 420$. **The lesser of the two page numbers facing Jenna is 20.**

(Note: Some advanced students may create and solve x(x +1) = 420 instead.)

FOLLOW-UP: The sum of the pages can be 420 but not 462. Why? [21 and 22 are not facing pages.]

3B. *Strategy: Find the largest number of quarters for one person.*
No one has more than 7 quarters since 8 quarters are worth $2.00. Since $1.85 ends in a 5 and
the value of dimes ends in a 0, each person has an odd number of quarters. Each person has a
different number of coins, so each has a different number of quarters. Then one of them has 1
quarter, one has 3 quarters, one has 5 quarters, and one has 7 quarters. $1 + 3 + 5 + 7 = 16$.
Together, they have 16 quarters.

*FOLLOW-UP: Tom is able to provide exact change if he buys any item whose cost is under $1.00. What is
the least number of coins he can have?* [9]

3C. **METHOD 1:** *Strategy: Examine the sum of any 2, 3, or 5 consecutive numbers.*
The sum of any two consecutive whole numbers is odd. The sum of any three consecutive
whole numbers is a multiple of 3◊. The sum of any five consecutive whole numbers is a multiple
of 5◊. Thus, the sum is an odd multiple of 15, the LCM(3,5). The least such multiple is 15.
Check that $15 = 7 + 8$, and also that $15 = 4 + 5 + 6$, and also $15 = 1 + 2 + 3 + 4 + 5$. **The least
whole number that can be expressed as the sum of 2, 3 and 5 whole numbers is 15.**

◊*Justification*: Start with the middle number, *M*. The sum of three consecutive whole numbers can be
written as $(M - 1) + (M) + (M + 1)$. This simplifies to 3 times the middle number *M*, which means the
sum is a multiple of 3. Similarly, the sum of five consecutive whole numbers can be shown to be 5 times
their middle number, which makes that sum a multiple of 5.

METHOD 2:   *Strategy: Examine the simplest cases using one condition first.*

Sum of 5 consecutive numbers	Sum of 2 consecutive numbers?	Sum of 3 consecutive numbers?
$0 + 1 + 2 + 3 + 4 = 10$	No	No
$1 + 2 + 3 + 4 + 5 = 15$	*Yes:* $7 + 8$	*Yes:* $4 + 5 + 6$

The least number that satisfies all 3 conditions is 15.

FOLLOW-UPS: *(1) What number times 201 equals the sum $1 + 2 + 3 + \ldots + 198 + 199 + 200$?* [100]

3D. METHOD 1:   *Strategy: Consider the sum of their distances.*

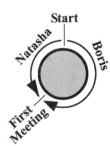

When Boris and Natasha first meet after starting, the sum of the distances they have run is equal to one complete lap (distance around the track). Every second, Boris runs 5 meters and Natasha runs 3 meters. Between them they cover 8 meters every second. They cover one complete lap in $200 \div 8 = 25$ sec. **They will first meet 25 seconds after starting.**

METHOD 2:   *Strategy: Use algebra.*

Suppose each runs for x seconds before they meet.
Then Natasha covers $3x$ and Boris $5x$ meters before they meet.

$$3x + 5x = 200$$
$$8x \quad = 200$$
$$x \quad = \ 25$$

They will first meet 25 seconds after starting.

FOLLOW-UPS: *(1) Suppose Boris starts running 4 seconds before Natasha. They run in opposite directions. How long will Natasha run before she meets Boris for the second time?* [47.5 sec] *(2) If Boris and Natasha start at the same time and same place and run in opposite directions, how long will it take for them to pass each other at the starting point?* [200 sec]

3E.   *Strategy: Express the area to be found in terms of figures whose areas are known.*

Connect the centers of the circles to form a square with side 4 cm. The area of the square is $4 \times 4 = 16$ cm². The unshaded region inside the square consists of 4 quarter-circles each of radius 2 cm. These 4 quarter-circles can be arranged to form a circle of radius 2 cm. The area of the circle is πr^2 or approximately $3.14 \times 2 \times 2$, which is 12.56.

The area of the shaded region can then be found by subtracting the area of the circle from the area of the square, $16 - 12.56 = 3.44$. Round this to the nearest tenth. **The area of the shaded region is 3.4 cm².**

4A. **METHOD 1A:** _Strategy: Find the interval starting from –11._
The starting point is –11 and the finishing point is –3. The total interval from –11 to –3 is $(-3) - (-11) = 8$. $\frac{3}{4}$ of 8 is 6. Then 6 more than –11 is –5. **The number that is $\frac{3}{4}$ of the way from –11 to –3 is –5.**

METHOD 1B: _Strategy: Find the interval starting from –3._
As above, –11 and –3 are 8 units apart. $\frac{3}{4}$ of the way from –11 gives the same result as $\frac{1}{4}$ of the way from –3. Then $\frac{1}{4}$ of 8 is 2, and 2 units less than –3 is –5.

METHOD 2: _Strategy: Draw a number line and count spaces._

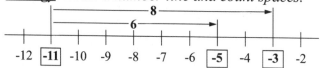

METHOD 3: _Strategy: Find successive averages._
Since half of a half is a fourth, average the endpoints twice. Halfway between –11 and –3 is –7. Halfway between –7 and –3 is –5. Therefore, $\frac{3}{4}$ of the way from –11 to –3 is –5.

4B. _Strategy: Use the LCM of 2, 3, 6, and 9._
Any number divisible by 2, 3, 6, and 9 is divisible by their least common multiple (LCM). The LCM of 9 and 6 is 18. The LCM of 9, 6, and 3 is still 18. The LCM of 9, 6, 3, and 2 is still 18. Then 33,822 is given as divisible by 18. **Increasing, the next number also divisible by 2, 3, 6, and 9 is** $33,822 + 18 = $ **33,840.**

FOLLOW-UPS: (1) What is the least number larger than 1 that leaves a remainder of 1 when divided by 10 or 12 or 18? [181] (2) What is the least number that leaves a remainder of 5 when divided by 6, a remainder of 7 when divided by 8, and a remainder of 9 when divided by 10? [119]

4C. _Strategy: Combine their results to get an equal number of lengths and widths._
Noelle's sum, 88 cm, is the result of adding two lengths and one width.
Ryan's sum, 80 cm, is the result of adding one length and two widths.

$$\text{Noelle} \quad W\begin{array}{|c|} \hline L \\ p = 88 \\ \hline L \end{array} \qquad \text{Ryan} \quad W\begin{array}{|c|} \hline L \\ p = 80 \\ \hline \end{array}W$$

If their sums are added together, the resulting 168 cm is the sum of three lengths and three widths. Then the sum of one length and one width (called the semiperimeter) is $168 \div 3 = 56$. Since the perimeter is the sum of two lengths and two widths, **the perimeter is** $2 \times 56 = $ **112 cm.**

4D. _Strategy: Consider the difference of their distances._
Because they start from the same place and run in the same direction, each time Boris and Natasha meet the _difference_ in the distances they run must be a whole number of laps (distances around the track). After they start, their first meeting occurs when Boris overtakes Natasha. He will have run 200 m farther. Each second Boris runs 2 m farther than Natasha. The elapsed time is $200 \div 2 = 100$ sec. At 3 mps, **Natasha has run 300 m when they first meet after starting.**

FOLLOW-UP: _Suppose on the same track Natasha starts directly opposite Boris. They run in the same direction. How far has Natasha run when Boris overtakes her?_ [150 m]

4E. **METHOD 1:** _Strategy: Combine the contents of the two balance scales._
Reverse the pans of the left balance scale. Then combine the contents of the left pans of the two scales, getting 2 triangles and a square. Combine the contents of the two right pans, getting 7 circles and a square. These resulting pans are in balance. After the square is removed from each pan, **7 circles balance two triangles.**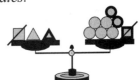

METHOD 2: _Strategy: Add two circles to each side of the left balance scale._
A. In the left balance scale, the left pan now has 7 circles and the right pan now has one triangle, one square, and two circles.

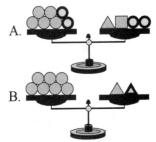

B. In the question the righthand diagram shows that one square and two circles weigh the same as one triangle. Replace the square and two circles on the right pan in A by one triangle. 7 circles balance two triangles.

METHOD 3: _Strategy: Add one triangle to each side of the right balance scale._
A. The right side of the balance now has two circles, one square, and one triangle.

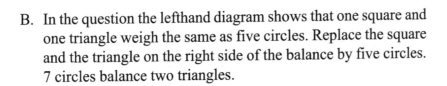

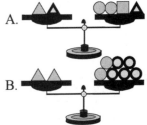

B. In the question the lefthand diagram shows that one square and one triangle weigh the same as five circles. Replace the square and the triangle on the right side of the balance by five circles. 7 circles balance two triangles.

5A. _Strategy: Feed all the cats first._
One quart of milk feeds 6 cats so 15 cats require $2\frac{1}{2}$ quarts of milk. This leaves $\frac{1}{2}$ quart of milk. One quart of milk feeds 10 kittens, so **5 kittens can be fed with the leftover milk**.

FOLLOW-UP: Mr. Brown can read 100 pages of a certain 300-page book in half the time it takes Mr. Green. They began reading the book at the same time and both stopped after two hours. If Mr. Brown was on page 150 when they stopped, how many more hours will it take Mr. Green to finish the book? [6 hours]

5B. **METHOD 1:** _Strategy: Find how many students passed the exam in each case._
75% of the 500 students = 375 students actually passed the exam. If only 10% had failed, then 90% of the 500 = 450 students would have passed. There would have been 450 – 375 = **75 more passing grades if only 10% had failed.**

METHOD 2: _Strategy: Work entirely with the percents of failure._
Since 75% of the students passed the test, 25% of them failed. If only 10% of them had failed, then 25% – 10% = 15% more students would have passed. 15% of 500 students is 75 students.

FOLLOW-UP: The average grade for 25 students on a certain test was 82. If 10 of the students had each earned 5 more points on their test, what would be the average grade for the test? [84]

5C. _Strategy: Find the least and greatest possible sums first._
The least sum = (–17) + (–16) = –33. The greatest sum = 17 + 16 = 33. Each integer from –33 to 33 inclusive is a possible sum. Then 33 negative sums, 33 positive sums, and zero are all possible. **The total number of different sums is 67.**

FOLLOW-UP: 10% of the sum of two integers is 12. One of the integers is negative. What is the least possible value of the positive integer? [121]

5D. **METHOD 1:** _Strategy: Split the figure into pieces of equal area._
Draw $\overline{GH}$ and $\overline{JK}$ parallel to $\overline{EF}$ so as to divide rectangle _EBCF_ into three regions of equal area. The area of _EBCF_ is 3 times the area of _AEFD_, and the total area is 144, so the area of each of the four small regions is 144 ÷ 4 = 36 cm². Because _AEFD_ is a square and the four small regions have equal bases and areas, all four regions are squares. The area of each square is 36 cm², so each side of a square is 6 cm. Then _EBCF_ is composed of 8 congruent segments each of length 6 cm, and **the perimeter of rectangle _EBCF_ is** 8 × 6 or **48 cm.**

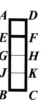

METHOD 2: *Strategy: Find the area of each region.*
The ratio of the area of rectangle *II* to the area of square *I* is 3:1. Then the area of rectangle *II* is $\frac{3}{4}$ of the area of rectangle *ABCD* and the area of square *I* is $\frac{1}{4}$ of the area of *ABCD*. The area of square *I* is 36 cm² and the area of rectangle *II* is 3×36 = 108 cm². Then *EF* = 6 cm and *EB* = 108 ÷ 6 = 18 cm. The perimeter of rectangle *II* is $(2 \times 18) + (2 \times 6)$, or 48 cm.

5E. **METHOD 1:** *Strategy: Count directly from 120 days ago to 86 days from now.*
86 days from today is 120 + 86 or 206 days after a Friday. A week is 7 days. 206 ÷ 7 = 29 R3. 29 weeks and 3 days after a Friday is the same day of the week as 3 days after the first Friday. **86 days from today is a Monday.**

METHOD 2: *Strategy: Count both ways from today.*
A week is 7 days. $7 \times 17 = 119$. Then 119 days ago was the same day of the week as today. 120 days ago was one day earlier in the week than today. If 120 days ago was a Friday, then today is a Saturday. Also, since $7 \times 12 = 84$ days from today is also a Saturday, 86 days from today is two days later, a Monday.

SET 16 ◆ ◆ ◆ **Olympiad 1**

1A. **METHOD 1:** *Strategy: Combine 2s and 5s to create a triangular array of 7s.*
The array of 5s may be seen as a mirror image of the array of 2s.
Reflect each 5 to the left and add it to its corresponding 2.
Adding columns, $42 + 35 + 28 + 21 + 14 + 7 = 147$.
The sum of the digits is 147.

```
7
7  7
7  7  7
7  7  7  7
7  7  7  7  7
7  7  7  7  7  7
```

METHOD 2: *Strategy: Separate the digits into an array of 2s and an array of 5s.*

```
2                              5
2  2                        5  5
2  2  2                  5  5  5
2  2  2  2            5  5  5  5
2  2  2  2  2      5  5  5  5  5
2  2  2  2  2  2   5  5  5  5  5  5
  21 × 2 = 42        21 × 5 = 105
```
The sum of the digits is 42 + 105 = 147.

(Note: An alternative is (17 × 2) + (4 × 7) + (17 × 5) = 147.)

1B. *Strategy: Find the time that each candle burns.*

Candle B burns at twice the rate of candle A, so candle B needs half the time that candle A does. Candle B burns for $72 \div 2$ or 36 minutes. Similarly, candle C burns one-third of the time that candle B does or 12 minutes. **The greatest number of minutes of light is** provided when the candles burn in succession, for $72 + 36 + 12$ or **120 minutes.**

1C. **METHOD 1:**   *Strategy: Write the second expression as two separate fractions.*

Rewrite $\frac{2003 + 25}{25}$ as $\frac{2003}{25} + \frac{25}{25} = \frac{2003}{25} + 1$.

Then compare $(\frac{2003}{25} + 25)$ and $(\frac{2003}{25} + 1)$. The difference is 24.
The first expression is greater than the second by 24.

METHOD 2:   *Strategy: Write each expression as a simple fraction.*

$\frac{2003}{25} + 25 = \frac{2003}{25} + \frac{625}{25} = \frac{2628}{25}$ and $\frac{2003 + 25}{25} = \frac{2028}{25}$.

Then $(\frac{2003}{25} + 25) - (\frac{2003 + 25}{25}) = \frac{2628}{25} - \frac{2028}{25} = \frac{600}{25} = 24$.
The first expression is greater than the second by 24.

1D. *Strategy: Make an exhaustive list.*

Three grey circles and two white circles can be arranged in the following orders:

All other arrangements can be obtained from these six by reversing their order (that is, by rotating the strip). **There are 6 different ways that three circles can be colored gray.**

FOLLOW-UPS: (1) Suppose just two circles are to be colored gray. In how many ways can this be done if the paper contains 2 circles? 4 circles? 6 circles? [1; 4; 9] (2) EXPLORATIONS: Find patterns by varying the number of circles and the number colored gray. Hint: Keep one number constant while varying the other, and then keep the second number constant while varying the first.

1E. **METHOD 1:**   *Strategy: Find the dimensions of the smaller rectangles.*

The perimeter of each small rectangle is 20 cm, so its semiperimeter (the sum of one length and one width) is 10 cm. The diagram shows that the length is 4 times the width. Then, for each small rectangle, the width is 2 cm, the length is 8 cm, and the area is $2 \times 8 = 16$ sq cm. **The area of the large rectangle is** $5 \times 16 = $ **80 sq cm.**

METHOD 2: *Strategy: Create a "unit" length.*
Assume the width of a small rectangle is 1 unit. Then the bottom of the larger rectangle is 4 units, as shown, as is its top. Thus each small rectangle is 1 unit by 4 units, with a perimeter of 10 units. Since the perimeter is given as 20 cm, each unit equals 2 cm. Therefore, each small rectangle measures 2 cm by 8 cm, and the larger rectangle is 8 cm by 10 cm, with an area of 80 cm.

FOLLOW-UP: *Five congruent rectangles each with whole number dimensions and perimeter 20 cm are arranged to form a larger rectangle. What is the greatest possible area of the large rectangle?* [125 sq cm] *Least area?* [45 sq cm] *Greatest perimeter?* [92 cm] *Smallest perimeter?* [28 cm]

2A. *Strategy: Substitute and simplify.*

$\frac{10 + 2 \times 3}{10 - 2 \times 3} = \frac{10 + 6}{10 - 6} = \frac{16}{4} = 4.$ **The value of** △ **is 4.**

FOLLOW-UPS: *(1) For what values of a, b, and c will the expression be undefined?* [If $a = b \times c$, the denominator is 0.] *(2) For what values of a, b, and c will the expression equal 1?* [Either b or c must be 0].

2B. **METHOD 1:** *Strategy: Compare each number to the average.*
19, 21, 24, and 25 are respectively 6, 4, 1, and 0 less than the average of 25. The sum of these differences is 11 less than the average. The fifth number must be 11 more than the average. **The other number is** $25 + 11 = \textbf{36.}$

METHOD 2: *Strategy: Compare the two totals.*
Five numbers that average 25 must have a total of $5 \times 25 = 125$. The sum of the given four numbers, 19, 21, 24, and 25, is 89. The other number is $125 - 89$ or 36.

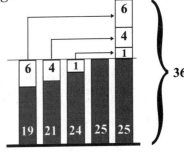

2C. **METHOD 1:** *Strategy: Find the prime factorization.*
Factor: $100 = 4 \times 25 = 2^2 \times 5^2$. Because 2^2 is a factor of 100, 2^0 and 2^1 are also factors of 100. Similarly, Because 5^2 is a factor of 100, 5^0 and 5^1 are also factors of 100. To get all the factors, set up the multiplication table shown. **There are 9 different whole numbers that are factors of 100.**

✗	$5^0 = 1$	$5^1 = 5$	$5^2 = 25$
$2^0 = 1$	1	5	25
$2^1 = 2$	2	10	50
$2^2 = 4$	4	20	100

(***Note:*** *To count the number of factors it is not necessary to do the multiplications. The table's 3 rows and 3 columns require $3 \times 3 = 9$ entries.*)

SOLUTIONS: DIVISION M

METHOD 2:   *Strategy: Use the Factor Pairs Table.*

100

1×100
2×50
4×25
5×20
10×10

There are 9 different whole numbers that are factors of 100.

FOLLOW-UPS: *(1) How many different whole numbers are factors of $2^7 \times 3^4$? [$8 \times 5 = 40$] (2) How many different whole numbers are factors of $11^3 \times 13^2 \times 17^1$? [$4 \times 3 \times 2 = 24$] (3) How many different whole numbers are factors of 1400? [24, as in FOLLOW-UP 2]*

2D. METHOD 1:   *Strategy: Find the least common multiple of the two cycles.*

TWENTY has 6 letters and will therefore be spelled correctly in every 6th line, starting with line 1. *FIVE* has 4 letters and will be spelled correctly in every 4th line. The LCM of 6 and 4 is 12. Then both words will be spelled correctly in line 1 and in every 12th line after that. The next line in which both are spelled correctly is line 13. **N is 13.**

1.	Twenty	Five
2.	yTwent	eFiv
3.	tyTwen	veFi
4.	ntyTwe	iveF
⋮	⋮	⋮
N.	Twenty	Five

METHOD 2:   *Strategy: List the lines in which each is spelled correctly.*

As above, *TWENTY* has 6 letters and is cycled every 6 lines, and
FIVE has 4 letters and is cycled every 4 lines.

TWENTY is spelled correctly in line 1 and in every 6th line after that: lines 1, 7, 13, 19, …
FIVE is spelled correctly in line 1 and in every 4th line after that: lines 1, 5, 9, 13, 17, …

The first match after line 1 is line 13, so $N = 13$.

FOLLOW-UPS: *(1) Find N if line N is the 6th time both words are spelled correctly. [61] (2) Suppose line 1 reads SIX HUNDRED TWENTY FIVE and each word is cycled separately as in the problem. What is the first line number greater than 500 for which all four words are again spelled correctly? [505]*

2E. METHOD 1:   *Strategy: Split the shaded region into simpler shapes.*

Extend the sides of the smaller square to split the shaded region into 2 congruent rectangles, each of area A, and an even smaller square of area B. The area of the shaded region, 28 sq cm, equals twice A plus B. Both 28 and twice A are even, so B is even. Then B is either 16 or 4. However, if B is 16, then A is 6, one side-length of the rectangle is 4, and the other side-length is not an integer. If B is 4, then A is 12, with side-lengths of 2 and 6. The side-length of the unshaded square is also 6. Thus the side-lengths of the large square is $6 + 2 = 8$ cm, and **its area is 64 sq cm.**

2E. **METHOD 2:** _Strategy: Use "Every square is the sum of consecutive odd integers 1, 3, 5, ..."_
The area of the large square can be written as $1 + 3 + 5 + \ldots +$ some odd integer. The area of the smaller square can be written in the same form. The difference in the areas is therefore the sum of some consecutive odd integers. Write 28 as $13 + 15$. The area of the smaller square is then $1 + 3 + 5 + 7 + 9 + 11$ or 36 sq cm and the area of the larger square is $1 + 3 + 5 + 7 + 9 + 11 + 13 + 15$ or 64 sq cm.

METHOD 3: _Strategy: Add 28 to possible areas of the small square._
Make an organized list of possible areas for the smaller square. Add the area of the shaded region, 28 sq cm, to each entry. Any sum that is a perfect square is the area of the larger square.

Side of small square	2	3	4	5	6	7
Area of small square	4	9	16	25	36	49
Add 28	32	37	44	53	64	77
Perfect square?	No	No	No	No	Yes	No

The area of the larger square is **64 sq in.**

3A. **METHOD 1:** _Strategy: Find the number of additional minutes._
The cost of the additional minutes is $40¢ - 25¢ = 15¢$. The number of additional minutes is then the $15¢$ cost divided by the $3¢$ cost per minute. Jason talks for 5 additional minutes so that **the call lasts for** $3 + 5 =$ **8 minutes.**

METHOD 2: _Strategy: Cost out the call minute by minute._
Starting with the cost of the first three minutes, $25¢$, add $3¢$ until the sum reaches $40¢$: $25 + 3 + 3 + 3 + 3 + 3 = 40¢$. Therefore, Jason's call lasts $3 + 5 = 8$ minutes.

METHOD 3: _Strategy: Use algebra._
Let $x =$ the number of additional minutes talked.
$25 + 3x = 40$; $3x = 15$; $x = 5$. Therefore, Jason's call lasts 8 minutes.

FOLLOW-UP: Phone Company A offers customers a plan where the first 3 minutes cost 25¢ and each additional minute costs 3¢. Phone Company B offers them a plan where the first three minutes cost 15¢ and each additional minute costs 4¢. For what length phone calls would Company A's plan be the better deal? [any time greater than 13 minutes]

3B. _Strategy: Find the factor pairs of 48._
List the pairs in a Factor Pairs table. Determine which pair has an average of 8. The average of 4 and 12 is 8. **The greater of the two numbers is 12.** A variation is to list pairs of whole numbers whose average is 8 and determine which pair has a product of 48: 8 and 8; 7 and 9; 6 and 10; 5 and 11; 4 and 12. The two numbers are 4 and 12, and the larger is 12.

Factors of 48		Average
1	48	(not whole)
2	24	13
3	16	(not whole)
4	12	8
6	8	7

3C. **METHOD 1:** _Strategy: Look for a pattern._
Let O represent an odd number and E an even number. The sequence begins $O, O, E, O, O, E, O, O, E, \ldots$ The table at the right shows all results of adding odd and even numbers. Use this table to show that the above pattern continues and that every third number in the sequence is even. Therefore, $\frac{1}{3}$ of the terms in the sequence are even, and $\frac{2}{3}$ are odd. Of the first 30 terms in the sequence, **20 are odd.**

O	+ O	=	E
E	+ E	=	E
E	+ O	=	O
O	+ E	=	O

METHOD 2: _Strategy: Write out the ones digits of the terms._
Whether a number is odd or even depends only on the ones digit. Continue the sequence to 30 terms, writing down the ones digits only.

1, 1, 2, 3, 5, 8, 3, 1, 4, 5, 9, 4, 3, 7, 0, 7, 7, 4, 1, 5, 6, 1, 7, 8, 5, 3, 8, 1, 9, 0.

Twenty of these numbers of these are odd.

METHOD 3: _Strategy: Write out the first 30 terms._
The sequence produces large numbers surprisingly quickly. The 30th term is 832,040. If the arithmetic is done correctly, 20 of the terms will be odd. This method is slow and error-prone.

FOLLOW-UPS: (1) How many of the first 200 terms are odd? [134] (2) How many of the first 2004 terms are multiples of 5? [400]

3D. _Strategy: Make a list of all possible lesser prime numbers._
If the sum of two numbers is 40, one number is 20 or less and the other is 20 or more. The first row lists the set of all possible lesser numbers which are also prime. The second row lists the difference between 40 and the lesser number. The third row notes which pairs are both prime. **40 can be expressed as the sum of two prime numbers in 3 ways.**

Prime p	2	3	5	7	11	13	17	19
$40 - p$	38	37	35	33	29	27	23	21
Is $40 - p$ prime?	No	Yes	No	No	Yes	No	Yes	No

FOLLOW-UPS: (1) Express 20 as the sum of two primes. [3 & 17, 7 & 13] (2) Find all pairs of primes whose sum is 100. [3 & 97, 11 & 89, 17 & 83, 29 & 71, 41 & 59, 47 & 53] (3) Find an even number that is not the sum of two primes. [2, which is the least prime] (4) Find the least prime number that is the sum of three different primes. [19]

(Note: A statement, known as the Goldbach Conjecture, says that every even integer greater than 2 can be expressed as the sum of two primes. Mathematicians have tried unsuccessfully for over 250 years to either prove or disprove the statement.)

SOLUTIONS: DIVISION M

3E. **METHOD 1:**   *Strategy: Use ratios.*
If $\overline{BA}$ and $\overline{BD}$ are used as bases, $\triangle CBA$ and $\triangle CBD$ share the same height. Then the ratio of their areas will be the same as the ratio of their bases. $BD = 19$ and $BA = 25$. The area of $\triangle CBD$ is $\frac{19}{25}$ of the area of $\triangle CBA$. $\frac{1}{2} \times 50 = 38$. **The area of $\triangle CBD$ is 38 sq cm.**

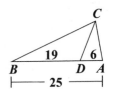

METHOD 2:   *Strategy: Use the formula for the area of a triangle.*
The area of a triangle is given by the formula $A = \frac{1}{2}bh$. The area of $\triangle CBA = 50$ sq cm, and base $BA = 25$ cm, so $50 = \frac{1}{2} \times 25 \times h$. Then $h = 4$ cm = the height of $\triangle CBA$. The height of $\triangle CBD$ is the same 4 cm if $\overline{BD}$ is used as the base. The area of $\triangle CBD$ is $\frac{1}{2} \times 19 \times 4 = 38$ sq cm.

4A. **METHOD 1:**   *Strategy: Build a table starting with the simplest numbers.*

Suppose:	HA – AH	Difference	H – A
$H = 2, A = 1$	21–12	9	1
$H = 3, A = 1$	31–13	18	2
$H = 4, A = 1$	41–14	27	3
$H = 3, A = 2$	32–23	9	1
$H = 4, A = 2$	42–24	18	2
$H = 5, A = 2$	52–25	27	3

The last two columns indicate that $HA - AH$ is 9 times as large as $H - A$. To see why this is true, see Method 2. Since $HA - AH = 18$, **then the value of $H - A$ is 2.**

METHOD 2:   *Strategy: Use algebra.*
A two-digit number can be represented by the sum of the ones digit and 10 times the tens digit. Therefore, the numbers HA and AH can be represented by $10H + A$ and $10A + H$, respectively. Their difference is $(10H + A) - (10A + H)$. This simplifies to $9H - 9A$ or $9(H - A)$. Thus, the difference between any two-digit number and its reversal is always a multiple of 9. In this case, since $9(H - A) = 18$, then $H - A = 2$.

FOLLOW-UP: Investigate the three-digit numbers FEB and BEF for similar patterns. [The difference between FEB and BEF is 99 times the difference between the values of F and B.]

4B. *Strategy: Start with an extreme case.*
If each dart scores 2 points, the minimum total score is 12. If each dart scores 8 points, the maximum total score is 48. Every 2 that is replaced by 5 or 8 increases the total score by 3 or 6 points. Therefore, the total score is any multiple of 3 between 12 and 48, inclusive. **Of the given scores only 36 is possible.** (It can be achieved, among other ways, by four 8s and two 2s.)

FOLLOW-UP: In how many ways can 36 points be scored? Why? [3. Scores of 8 and 2 can be replaced by scores of 5 and 5. 36 is gained 3 ways: $4 \times 8 + 2 \times 2$; $3 \times 8 + 2 \times 5 + 2$; $2 \times 8 + 4 \times 5$.]

SOLUTIONS: DIVISION M

4C. **METHOD 1:** *Strategy: Find a pattern.*
The multiples of 4 occur in either column A or D. However, the multiples of 4 in column A are also multiples of 8. Since 100 is a multiple of 4, but not of 8, **100 will appear in column D.**

METHOD 2: *Strategy: Unfold the columns so that they appear as shown.*
Split the table, odd-numbered rows on the left and even-numbered rows, reversed, on the right. The multiples of 8 will appear in column A2.

A1	B1	C1	D1	D2	C2	B2	A2
1	2	3	4	5	6	7	8
9	10	11	12	13	14	15	16
⋮	⋮	⋮	⋮	⋮	⋮	⋮	⋮

Therefore, 96 appears in A2 and 100 appears in D1. 100 appears in column D.

4D. **METHOD 1:** *Strategy: Represent percents as fractions.*
$75\% = \frac{3}{4}$ and $\frac{3}{4}$ of 76 is 57. $95\% = \frac{19}{20}$ and if $\frac{19}{20}$ of a number is 57, then $\frac{1}{20}$ of the number is 3, and $\frac{20}{20}$ of the number is 60. **The whole number is 60.**

METHOD 2: *Strategy: Use algebra and decimals.*
If N is the number to be found, then 75% of 76 is 95% of N.
$$0.75 \times 76 = 0.95N$$
$$57 = 0.95N$$
$$N = 57 \div 0.95$$
$$N = 60 \quad \text{The whole number is 60.}$$

METHOD 3: *Strategy: Use algebra and proportions.*
Let N be the number to be found and x be 75% of 76.

(1) $\frac{75}{100} = \frac{x}{76}$ (Reduce $\frac{75}{100}$)

    $\frac{3}{4} = \frac{x}{76}$ (Clear fractions)

    $4x = 228$ (Divide by 4)

    $x = 57$ (Substitute into 2nd proportion)

(2) $\frac{95}{100} = \frac{57}{N}$ (Reduce $\frac{95}{100}$)

    $\frac{19}{20} = \frac{57}{N}$ (Clear fractions)

    $19N = 1140$ (Divide by 19)

    $N = 60$ The whole number is 60.

FOLLOW-UPS: *(1) 40% of 60 is what percent of 72? [$33\frac{1}{3}$%] (2) 25% of what number is 120% of 15? [72]*

4E. *Strategy: Maximize the number of large tiles.*
For a fixed area, maximize the number of large tiles to minimize the total number of tiles. The area of the floor is 99 sq ft and the area of each 2 × 3 ft tile is 6 sq ft. Then the floor is covered by 16 large tiles with 3 sq ft left over, because $99 \div 6 = 16$ R3. Thus, 3 small tiles are needed. It is also necessary to prove that 19 tiles *can* cover the floor; the diagram shows one way to do this. **The least total number of tiles is 19.**

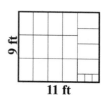

FOLLOW-UPS: *(1) Suppose the rectangular floor is 7 ft by 11 ft. What is the minimum number of 2 × 3 and 1 × 1 tiles needed to cover the floor? [17] (2) Suppose the floor is 9 ft by 11 ft. What is the minimum number of 2 × 3 and 5 × 1 tiles needed to cover the floor? [17. Show how this can be done.]*

SOLUTIONS: DIVISION M

5A. **METHOD 1:**   *Strategy: Compare what is given with what is sought.*

Whatever number "⌂" represents, just add 50 to $3 \times ⌂ - 25$ to get $3 \times ⌂ + 25$. Then just add the same 50 to the 8: **the value of $3 \times ⌂ + 25$ is 58.**

METHOD 2:   *Strategy: Use algebra.*

$$3 \times ⌂ - 25 \ = 8 \quad \text{(Add 25)}$$
$$3 \times ⌂ \qquad = 33 \quad \text{(Divide by 3)}$$
$$⌂ \qquad = 11 \quad \text{(Substitute)}$$

Since $3 \times 11 + 25 = 33 + 25$, the value of $3 \times ⌂ + 25$ is 58.

5B. **METHOD 1:**   *Strategy: Square the multiples of 7.*

If the square of a number is a multiple of a prime, the number itself must be a multiple of that prime. Of $7^2 = 49$, $14^2 = 196$, and $21^2 = 441$, only 14^2 is between 100 and 200. **The whole number between 100 and 200 that is both a perfect square and a multiple of 7 is 196.**

METHOD 2:   *Strategy: List the perfect squares between 100 and 200.*

$11^2 = 121$, $12^2 = 144$, $13^2 = 169$, and $14^2 = 196$.

Of these, only 196 is a multiple of 7. The desired number is 196.

FOLLOW-UPS: (1) Explain why a perfect square has an odd number of factors. [Factors of a number can be written in pairs, producing an even number of factors. The only exception is a number which is a perfect square. In that case, one factor will be paired with itself, producing an odd number of factors.] (2) How many even numbers less than 400 have an odd number of factors? [9]

5C. *Strategy: Use the Triangle Inequality.*

The sum of the lengths of two sides of a triangle must be greater than the length of the third side.

    (1) Suppose N is the longest side-length of the triangle.

        Then $6 + N > 25$, so $N > 19$. Rewrite as $19 < N$

    (2) Suppose N is one of the other two sides-lengths of the triangle.

        Then $6 + 25 > N$, so $31 > N$. Rewrite as $N < 31$

So, $19 < N < 31$. That is, $N = 20, 21, 22, 23, 24, 25, 26, 27, 28, 29,$ or 30.

11 whole numbers are possible values of N.

FOLLOW-UP: How many non-congruent triangles have a perimeter of 12, if the length of each side is a whole number? [3]

5D. *Strategy: Work backwards.*

After the exchange, Maria has 3 marbles for every 1 that Juan has, a ratio of 3:1. That is, of the 72 marbles Maria has 3 of every 4 and Juan has 1 of every 4. Therefore, Maria ends with 54 and Juan with 18 marbles.

	Total	**Juan**	**Maria**
Step 3. Juan has $\frac{1}{4}$ of the marbles.	72	18	54
Step 2. Juan gives Maria 12 marbles.	72	30	42
Step 1. Juan gives Maria $\frac{1}{2}$ his marbles.	72	60	**12**

Before Juan gave 12 of his marbles to Maria, she had $54 - 12 = 42$ and Juan had $18 + 12 = 30$. Before Juan gave half of his marbles to Maria, he had $2 \times 30 = 60$ marbles and she had $42 - 30 = 12$ marbles. **Maria originally had 12 marbles.**

5E. *Strategy: Make an organized list.*

For each numerator from 2 through 9, list all fractions whose value is between $\frac{1}{2}$ and 1. (Alternatively, make a list for all possible denominators, 3 through 10.)

Numerator	**Fractions**				**# of ways**
2	$\frac{2}{3}$				1
3	$\frac{3}{4}$	$\frac{3}{5}$			2
4	$\frac{4}{5}$	$\frac{4}{6}$	$\frac{4}{7}$		3
5	$\frac{5}{6}$	$\frac{5}{7}$	$\frac{5}{8}$	$\frac{5}{9}$	4
6	$\frac{6}{7}$	$\frac{6}{8}$	$\frac{6}{9}$	$\frac{6}{10}$	4
7	$\frac{7}{8}$	$\frac{7}{9}$	$\frac{7}{10}$		3
8	$\frac{8}{9}$	$\frac{8}{10}$			2
9	$\frac{9}{10}$				1
				TOTAL =	20

There are 20 different ways to get a value greater than $\frac{1}{2}$ and less than 1.

FOLLOW-UPS: (1) As the numerator increases from 2 to 5, why does the number of fractions increase by one? (2) As the numerator increases from 6 to 10, why does the number of fractions decrease by one? (3) Suppose the whole numbers that a and b are chosen from are 1, 2, 3, ... , 20. In how many different ways will $\frac{a}{b}$ be between $\frac{1}{2}$ and 1? [90] (4) How can the number of ways be found without listing?

1A. _Strategy: Find the weight of the tacks._
The difference, 120 – 70, represents half the weight of the tacks. The tacks in a full box weigh 100 g, so **the box weighs** 120 – 100 = **20 g when empty.**

1B. **METHOD 1:** _Strategy: Count the pairs that each member belongs to._
Call the members A, B, C, D, E, and F. A can be paired with each of the others, a total of 5 pairs containing A. Similarly, there are 5 pairs containing each of the other members. This would give a list of 6×5 or 30 pairs. This list, however, contains each pair twice. For example AB is listed as a pair containing A and again as a pair containing B. There are therefore $30 \div 2$ or 15 distinct pairs. **The greatest number of school days that can pass without repeating the same pair of students is 15.**

METHOD 2: _Strategy: List the pairs in an organized way._
For example, the pairs could be written as:

AB	AC	AD	AE	AF
	BC	BD	BE	BF
		CD	CE	CF
			DE	DF
				EF

There are $5 + 4 + 3 + 2 + 1$ or 15 pairs and therefore 15 days that can pass without repeating the same pair of students.

FOLLOW-UPS: (1) Six people shake hands with each other. How many handshakes take place all together? [15; this is really the same problem as 1B!] (2) What is the total number of diagonals that can be drawn in a polygon of 8 sides? [20] (3) N points on a circle determine 28 chords. What is the least value of N? [8] (4) A committee has 21 members, but only 20 serve at any one time. What is the maximum number of times the committee can meet before it must repeat a group of 20? [21; since only one is absent at a time, each member can be absent once.]

1C. _Strategy: Narrow the possibilities, one condition at a time._

Condition	Possible values of AB
1. $B \times A$ ends in an 8.	18, 81; 24, 42; 29, 92; 36, 63; 47, 74; 68, 86
2. 1st partial product has 2 digits.	18, 24, 29
3. 2nd partial product has 2 digits.	24

$$
\begin{array}{r}
A\ B \\
\times\ \ B\ A \\
\hline
\square\ 8 \\
\square\ \square \\
\hline
\square\ \square\ \square\ \square
\end{array}
$$

1st partial product ⟶
2nd partial product ⟶

Then **the four-digit product is 1008** because $24 \times 42 = 1008$.

1D. *Strategy: Draw a picture.*

As the circle rolls around the inside of the rectangle, its center traces out the perimeter of a smaller rectangle inside the larger one (see figure). Since the center is always 1 cm from the side of the larger rectangle, the smaller rectangle is 4 cm wide by 8 cm long. Hence **the center of the circle travels 24 cm.**

FOLLOW-UPS: (1) If the rectangle is a square 20 m on a side and the diameter of the circle is 3 m, what distance would now be traced by the center of the circle rolling around the interior of the square? [68 m] (2) Using the original measurements, suppose the circle rolled around the _outside_ of the rectangle. Why would the path of the center not be a rectangle? [As the circle pivots around each vertex of the rectangle, the center traces out a quarter-circle.] (3) In this case what is the distance traveled by the center of the circle? [$32 + 2\pi$ cm]

1E. *Strategy: Determine the fractional part of a lap covered by each between meetings.*

Sadie requires 45 seconds to go around the track. In the 30 seconds between two successive meetings, Sadie completes $\frac{30}{45} = \frac{2}{3}$ of a lap. Then the diagram shows that Rose completes $\frac{1}{3}$ of a lap in those 30 seconds. Therefore **Rose takes** 3×30 or **90 seconds to complete one lap.**

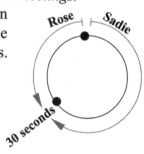

FOLLOW-UP: Suppose Sadie needs 35 seconds to complete one lap and Rose needs 40 seconds to complete one lap. If they both start from the same point, go in the same direction, and Sadie covers exactly one lap more than Rose does in the same time, how many seconds does Sadie require to run exactly one more lap than Rose? [280 seconds]

SET 17   Olympiad 2

2A. **METHOD 1:** *Strategy: Use the standard multiplication algorithm.*

We could employ the algorithm fully, but it is faster and more efficient to use only those parts we need. When the product is divided by $63 + 37 = 100$, the remainder will be the two rightmost digits of the product. Because the numbers being multiplied have zeros in the tens places, the two rightmost digits of the product are 21. **The remainder is 21.**

$$
\begin{array}{r}
8\ 0\ 3 \\
\times\quad 9\ 0\ 7 \\
\hline
-\ -\ 2\ 1 \\
-\ -\ 0 \\
\hline
-\ -\ -\ - \\
\hline
-\ -\ -\ -\ 2\ 1
\end{array}
$$

METHOD 2:   *Strategy: Divide first.*
Each of $ab \div 100$, $(\frac{a}{100}) \times b$, and $a \times (\frac{b}{100})$ produce the same result. Therefore, divide either 803 or 907 by 100. The remainder will be the digits in the decimal portion of the product of 8.03 and 907 or of 803 and 9.07. The remainder is 21 because $3 \times .07 = .03 \times 7 = .21$.

FOLLOW-UPS: *(1) What is the remainder when the product* 803×907 *is divided by 5?* [1] *By 20?* [1] *By 25?* [21] *(2) Without actually doing the division, can you find the remainder when 2345 is divided by 33?* [Dividing by 99 leaves a remainder of 1 for every hundred. $23 + 45 = 68$. The remainder is 2.]

2B. **METHOD 1:**   *Strategy: Build up from any four consecutive even integers.*
If the average of four numbers is 17, the sum of those numbers is $4 \times 17 = 68$. Take any four consecutive even integers, say, 2, 4, 6, and 8. Their sum is 20. To increase the sum by 48 while keeping the common difference of 2, increase each number by $48 \div 4 = 12$. The integers are then $2 + 12$, $4 + 12$, $6 + 12$, and $8 + 12$, or 14, 16, 18, and 20. **The greatest of the 4 integers is 20.**

METHOD 2:   *Strategy: Count outward from the average.*
Average is defined as a measure of central tendency. The four even integers, being consecutive, center around 17. Thus, the four even integers, building outward, are 16 and 18, and 14 and 20. Their average is 17, and the greatest of them is 20.

METHOD 3:   *Strategy: Use algebra.*
Let x = the least of the consecutive even integers. The others are $x + 2$, $x + 4$, and $x + 6$. Then $[x + (x + 2) + (x + 4) + (x + 6)] \div 4 = 17$. Solve to get $x = 14$. The integers are 14, 16, 18, and 20. The greatest is 20.

FOLLOW-UPS: *(1) The average of 5 different numbers is 23. If each number is increased by 6, what is their average now?* [29] *(2) The average of two numbers is 12 and the average of three other numbers is 22. What is the overall average of the set of 5 numbers?* [18]

2C. **METHOD 1:**   *Strategy: Use the test of divisibility for 8.*
Since 5 is the remainder, $3456n2$ is a multiple of 8. If a number is divisible by 8, then the number formed by its last 3 digits is also divisible by 8*. Thus, $6n2$ is divisible by 8. Since 600 is divisible by 8, then the two-digit number $n2$ is divisible by 8. The last two digits are 32 or 72, and **n is 3 or 7.** (The original number can be 345637 or 345677.)
   **Proof:* *Write 3456n2 as $345,000 + 6n2$. Note that $1000 = 8 \times 125$. Then both 345,000 and 3456n2 are multiples of 8. If any two terms of $3456n2 = 345,000 + 6n2$ are divisible by 8, then the third term is also divisible by 8 (distributive law). Hence, 6n2 is divisible by 8.*

METHOD 2:   *Strategy: Perform the division.*
Divide 3456 by 8. Because the remainder is zero, the last two digits $n2$ form a multiple of 8. Proceed as in method 1.

FOLLOW-UPS: *(1) Why do the two values of 3456n2 differ by 40?* [Because both end in 2, their difference is a multiple of 10. Both are also multiples of 8. Thus the difference is a multiple of 40.] *(2) For what values of H is 234H52 divisible by 8?* [1, 3, 5, 7, 9] *(3) 23A6 is divisible by 24. Find A.* [7]

2D. _Strategy: Draw a picture._

There are two possibilities. In the first picture, the 10 m side has moved right. To enclose an extra 40 sq m, the two 20 m sides are each lengthened by 4 m. This requires an extra 8 m of fencing.

In the second picture, the 20 m side has moved down. To enclose an extra 40 sq m, the two 10 m sides are each lengthened by 2 m. This requires an extra 4 m of fencing. **The least number of meters of additional fencing needed is 4 m.**

2E. _Strategy: Determine which numbers do <u>not</u> belong on the list._

        The perfect squares: 1, 4, 9, 16, 25, 36, 49, 64, 81, 100, . . .
        The perfect cubes: 1, 8, 27, 64, 125, . . .

Suppose the natural numbers from 1 through 75 are listed. Go through the list, and each time a square or cube is found, delete it and add the next consecutive integer at the end of the list. In this way the list will always contain 75 numbers.

Delete	1	4	8	9	16	25	27	36	49	64	**81**
Add on	76	77	78	79	80	**81**	82	83	84	85	**86**

Note that 81 is added on and later deleted. **The 75ᵗʰ term of the sequence is 86.**

Follow-Ups: (1) Suppose squares of perfect squares (that is, fourth powers) were also eliminated from the list. How would that affect the answer? [It wouldn't.] (2) What would the 75ᵗʰ number be if fifth powers were also eliminated? [87]

SET 17   Olympiad 3

3A. _Strategy: Use the process of elimination._

Neither A nor B can be 1 because the 1 is on the (hidden) face opposite the 6. This leaves 2, 3, 4, and 5 as possibilities for A and B. **The least possible value of $A + B$ is $2 + 3$, or 5.**

Follow-Up: The whole numbers 3, 4, 8, 9, 13, 14 appear on the six faces of a cube. Every pair of opposite faces has the same sum. What is that sum? [17]

3B. **METHOD 1:**   _Strategy: Change Curt's result to the correct result._

Multiplying Curt's result by 0.1 produces the original number and then multiplying the original number by 0.1 produces the correct result. Equivalently, multiplying Curt's result by .01 produces the correct result, and it has two decimal places. Since their difference is 33.66, the Curt's result is an integer, 34, the correct result is 0.34, and $34 - 0.34 = 33.66$. Thus, **the original positive number is 3.4.**

METHOD 2: *Strategy: Use algebra.*
Let N represent the original number.
$10N - .1N = 33.66$
$9.9N = 33.66$. Divide both sides of the equation by 9.9 to get $N = 3.4$, the original number.

3C. **METHOD 1:** *Strategy: Count the number of multiples of 3, 5, and 15 in the given interval.* The interval from 1 through 50 contains 16 multiples of three, of which 3 are in the interval from 1 through 10. Then there are 13 multiples of three in the given set. The interval from 1 through 50 contains 10 multiples of five, of which 2 are in the interval from 1 through 10. Then there are 8 multiples of five in the given set.

However, any number which is a multiple of both 3 and 5 is also a multiple of 15. There are 3 multiples of 15 in the given set, each of which has been counted twice.

The total number of values divisible by 3 or 5 is then $13 + 8 - 3 = 18$. **The probability that a number chosen from the interval is divisible by 3 or 5 is $\frac{18}{40}$ or $\frac{9}{20}$ or .45 or 45%.**

METHOD 2: *Strategy: List the numbers in a Venn Diagram.* There are 18 different numbers listed. Then the probability is $\frac{18}{40}$ or $\frac{9}{20}$ or .45 or 45%.

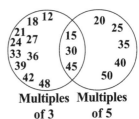

Multiples of 3   Multiples of 5

FOLLOW-UPS: (1) Given any 40 consecutive natural numbers, what is the least possible number of them that are divisible by 3 or 5? [18] The greatest? [20] (2) How many whole numbers between 30 and 100 are either a multiple of 7 or contain the digit 7? [24]

3D. *Strategy: Examine the difference in the perimeters of the tower and the large square.* Since $AB = DC$, the difference in the two given perimeters, 52 cm − 40 cm = 12 cm, is the sum of AD and BC. However, $AD = BC$, so $AD = 6$. Then the **perimeter of the small square is** 4×6 cm = **24 cm.**

FOLLOW-UPS: (1) What is the ratio of the area of the large square to the area of the tower? [25:34] (2) Suppose a small equilateral triangle of perimeter 15 cm is attached to each of the four sides of the large square. Find the perimeter of the resulting figure. [60 cm; you may wish to use this question to introduce fractals.] (3) What is the ratio of the areas of two squares if the ratio of their perimeters is 1:2? 1:5? 3:4? 11:6? [1:4; 1:25; 9:16; 121:36]

3E. **METHOD 1:** *Strategy: Write the fractions with a common numerator.* The least common multiple of the numerators is 12. Write each fraction with a numerator of 12: $\frac{12}{320} < \frac{12}{12n} < \frac{12}{303}$. The only multiple of 12 between 303 and 320 is 312, which is 12×26. **The whole number value of n is 26.**

METHOD 2:   *Strategy: Find the multiplicative inverses.*

If a and b are two positive numbers and $a < b$, then $\frac{1}{a} > \frac{1}{b}$. That is, reciprocals of unequal positive numbers are unequal <u>in the opposite order</u>. In this case, since $\frac{3}{80} < \frac{1}{n} < \frac{4}{101}$, then $\frac{80}{3} > \frac{n}{1} > \frac{101}{4}$. Since n is a whole number between 26.67 and 25.25, the value of n is 26.

METHOD 3:   *Strategy: Estimate the given fractions using simpler fractions.*

$\frac{3}{81}$, which equals $\frac{1}{27}$, is less than $\frac{3}{80}$.

$\frac{4}{101}$ is less than $\frac{4}{100}$, which equals $\frac{1}{25}$.

Thus, $\frac{1}{n}$ is between $\frac{1}{25}$ and $\frac{1}{27}$, and the value of n is 26.

SET 17   **Olympiad 4**

4A. **METHOD 1:**   *Strategy: Use the distributive property.*

By the (extended) distributive property, $ba + ca + da = (b + c + d)a$

Then $(17 \times 13) + (61 \times 13) + (22 \times 13) = (17 + 61 + 22) \times 13$.

The value is $(100) \times 13$ or **1300**.

METHOD 2:   *Strategy: Do the arithmetic .*

$(17 \times 13) + (61 \times 13) + (22 \times 13) = 221 + 793 + 286$. The value is 1300.

FOLLOW-UP: *Evaluate $(66 \times 28 + 44 \times 39 - 10 \times 27) - (66 \times 25 + 44 \times 36 - 10 \times 24)$* [300]

4B. *Strategy: Draw the possible diagrams.*

There are two possibilities. The upper diagram shows that two interior fences are needed, each of length 6 m, for a total of 12 m. The lower diagram shows that one interior fence of 6 m is needed along with one of 3 m, for a total of 9 m. **The least number of meters of additional fencing needed is 9 m.**

4C. **METHOD 1:**   *Strategy: Use the process of elimination.*

For one mode to exist, A must represent one of the numbers already in the list. The median is neither the greatest nor the least number. Then both the median and A must be one of the middle numbers, either 1.7 or 1.8. Suppose A is 1.7. Then the mode and median are both 1.7. The mean is $(1.6 + 1.7 + 1.7 + 1.8 + 2.1) \div 5 = 1.78$, not 1.7. Suppose A is 1.8. Then the mode and median are both 1.8. The mean is $(1.6 + 1.7 + 1.8 + 1.8 + 2.1) \div 5 = 1.8$. Since the mean, median, and mode are all equal, A **represents 1.8**.

METHOD 2: *Strategy: Use means, sums, and mode to build an equation.*
The sum of all five numbers is $7.2 + A$. Since all four given numbers are different, A is the mode of the five numbers. Then A is also the mean of the five numbers, whose sum therefore is $5A$. Equating sums, $5A = 7.2 + A$, and $A = 1.8$.

4D. **METHOD 1:** *Strategy: Use the properties of factors.*
Let N represent any whole number that has exactly three factors. Two factors of N are 1 and N. The third factor cannot be 1 nor N. Call it F.

To produce a product of N, F must be multiplied by one of the three factors: 1, N, or F. But F cannot be multiplied by either 1 or N, because in either case the result would not be N. The only remaining possibility is that F must be multiplied by itself, so that $F \times F = N$. Thus, N is the square of F.

What sort of number is F? Any factor of F is also a factor of N. Then F must be a prime number, or else N would have additional factors. Thus, N is the square of a prime number. **The counting numbers less than 50 with exactly three factors are 4, 9, 25, 49.**

METHOD 2: *Strategy: Check the perfect squares less than 50.*
Suppose all possible factor pairs of a given number are written. If all of the factors are different, there will be an even number of factors. If two of the factors are the same, — i.e. if the given number is a perfect square — there will be an odd number of factors. Therefore, only perfect squares have an odd number of factors. Make a list of the factors of perfect squares.

Number	1	4	9	16	25	36	49
Factors	1	1, 2, 4	1, 3, 9	1, 2, 4, 8, 16	1, 5, 25	1, 2, 3, 4, 6, 9, 12, 18, 36	1, 7, 49

The counting numbers that have exactly three factors are 4, 9, 25, and 49.

*FOLLOW-UP: "The Locker Problem" — One student runs through the school opening all lockers. A second student closes every second locker. A third student **changes** every third locker, (closing the open lockers and opening the closed lockers). A fourth student changes every fourth locker and so on. If 100 students follow this pattern, how many of the first 100 lockers in the school will be open? [10: only those that were changed an **odd** number of times.]*

4E. **METHOD 1:** *Strategy: Consider the total elapsed time.*
Start 5 years ago. At that time, Lucy's age was a multiple of Max's. Seven years later, (2 years from now), Lucy's age is again a multiple of Max's. If after adding 7, a prime number, to the two numbers, the second is still a multiple of the first, it strongly suggests that the original numbers are both multiples of 7. (This can be proved algebraically). Suppose five years ago Max was 7 years old. At that time Lucy was $3 \times 7 = 21$. Two years from now, Max will be $7 + 7 = 14$, and Lucy will be $21 + 7 = 28$, twice as old as Max. Thus, **Max is 12 years old now.**

SOLUTIONS: DIVISION M

4E. **METHOD 2:** *Strategy: Make a chart.*

Ages 5 years ago: Max, Lucy	4, 12	5, 15	6, 18	7, 21	8, 24
Ages in 2 years: Max, Lucy	11, 19	12, 22	13, 25	14, 28	15, 31
Is Lucy's age twice Max's?	No	No	No	Yes	No

Five years ago Max was 7 years old. Max is now 12 years old.

METHOD 3: *Strategy: Use algebra.*

5 years ago Lucy was 3 times as old as Max:
In two years both will be 7 years older:
In two years Lucy will be 2 times as old as Max:
Lucy's age is expressed two different ways:
Solve to get $M = 7$, Max's age 5 years ago. Max is 12 years old now.

	Max's age	Lucy's age
5 Years ago	M	$3M$
2 years from now	$M + 7$	$3M + 7$
2 years from now	$M + 7$	$2(M + 7)$

$$3M + 7 = 2(M + 7)$$

FOLLOW-UPS: *(1) Michael was born in the year 1992. His Mom was born in the year 1956. In what year will Michael's Mom be $2\frac{1}{2}$ times as old as Michael?* [2016. Thanks to PICO Kathy Bechard and her sons Michael and Casey who submitted this question. Students should be encouraged to create follow-up problems of their own.] *(2) In September, a club contains twice as many boys as girls. In October, 18 more girls join so that the club contains twice as many girls as boys. How many boys were in the club in September?* [12]

5A. *Strategy: Use the tests of divisibility.*

First, 52, 54, 56, and 58 are multiples of 2. Next, 51, 54, and 57 are multiples of 3 because the sum of the digits of each is also divisible by 3. Then, 55 is a multiple of 5. That leaves 53 and 59, both of which are prime. **The sum of the prime numbers between 50 and 60 is 112.**

FOLLOW-UPS: *(1) What is the least prime number that can be expressed as the sum of two prime numbers?* [5 = 2+3; 1 is neither prime nor composite] *three prime numbers?* [7 = 2+2+3] *(2) How many primes less than 20 can be expressed as the sum of three primes? How many ways?* [5 primes, 12 ways: **7** = 2+2+3; **11** = 2 + 2 + 7 = 3 + 3 + 5; **13** = 3 + 3 + 7 = 3 + 5 + 5; **17** = 2 + 2 + 13 = 3 + 3 + 11 = 3 + 7 + 7 = 5 + 5 + 7; **19** = 3 + 3 + 13 = 3 + 5 + 11 = 5 + 7 + 7.]

5B. *Strategy: Work backwards, using inverse operations.*

Question: [3] ← sq. rt. [?] ← +1 [?] ← sq. rt. [?] ← −6 [?] ← ÷2 [?] START

Solution: [3] → square [9] → −1 [8] → square [64] → +6 [70] → ×2 [140]

Adnan's original number was 140.

5C. _Strategy: Draw a picture and assign convenient lengths to segments._

Choose a unit of length for $\overline{BC}$, say, 1 cm. Then $CD = 1$ cm, $BD = 2$ cm, and $AB = 3$ cm. Since $AD = 5$ cm, $DE = 5$ cm. Thus $AE = 10$ cm, $AC = 4$ cm, and **AC is 40% of AE.**

FOLLOW-UPS: _(1) Suppose $AC = 9$. What is the length of $\overline{DE}$?_ [11.25] _(2) Suppose $\overline{BA}$ is 7 meters longer than $\overline{DB}$. What is the length of $\overline{EA}$?_ [70 m] _(3) What percent of BA is EB?_ [$233\frac{1}{3}$%]

5D. **METHOD 1:** _Strategy: Find a common factor for the 2 terms of the expression._
Represent the consecutive whole numbers by a and b, with $a < b$ and n as the final result: then $n = ab + 17$. According to the distributive property, if 17 is added to (or subtracted from) a multiple of 17, the result is another multiple of 17. For example, $99 \times 17 + 17$ yields 100×17. The number n is not prime if either $a = 17$ or $b = 17$. If $a = 17$, then $b = 18$; if $b = 17$, then $a = 16$. In both cases the results ($17 \times 18 + 17$ and $16 \times 17 + 17$) are multiples of 17. **The two values of the lesser of the consecutive numbers are 16 and 17.** The wording of the question tells us that no other possibilities less than 20 exist.

METHOD 2: _Strategy: List and test the possible values._

Lesser Consecutive Number	4	5	6	...	16	17	18	19
Result	37	47	59	...	289	323	359	397
Prime?	P	P	P	All Prime	17×17	17×19	P	P

The two values of the lesser of the consecutive numbers are 16 and 17.

5E. _Strategy: List the times and look for a pattern._
List the times by hour, beginning with 8:00, during which the HH is greater than MM.
Hour 8: 8:00, 8:01, 8:02, ... , 8:07   Total = 8 minutes
Hour 9: 9:00, 9:01, 9:02, ... , 9:08   Total = 9 minutes.
During each hour, the number of minutes when HH is greater than MM is the same as HH.
Combining $8 + 9 + 10 + 11 + 12 + 1 + 2$, **HH is greater than MM for a total of 53 minutes.**

FOLLOW-UPS: _Military time, based on a 24 hour day with no AM or PM, is written in the form HHMM from 0000 to 2359. PM is represented by adding 1200 to the common time. Thus 9:30 AM is written as 0930 and 9:30 PM is written as 2130. (1) Use military time to answer question 5E._ [77] _(2) In military time, what is the greatest number of minutes between consecutive palindromes?_ [251 minutes; it happens twice.] _(3) How many times during a 24-hour period is the military time a four-digit palindrome?_ [16; from 0110 to 0550, from 1001 to 1551 and from 2002 to 2442.]

SOLUTIONS: DIVISION M

ALSO FROM MOEMS

PUBLICATIONS & CONTESTS

MATH OLYMPIAD CONTEST PROBLEMS
FOR ELEMENTARY AND MIDDLE SCHOOLS
by Dr. George Lenchner.

SPECIFICATIONS: *Paperback. 8.5 in by 11 in. 280 pages. 1997 edition. ISBN 0-9626662-1-1. Library of Congress Catalog Number: 96-77380. Current prices are available at our Web site, www.moems.org. Add charges for shipping and handling.*

The predecessor to Volume 2, Dr. Lenchner's book is a collection of another 400 stimulating and challenging problems, complete with hints and detailed solutions. Compiled from the first 16 years of Olympiad contests, the 80 Olympiads offer an unusual variety of problems for young students, their teachers, and parents. These problems and seven unique appendices extend and enrich the elementary and middle school mathematics curriculum.

CREATIVE PROBLEM SOLVING
IN SCHOOL MATHEMATICS, 2ND EDITION
by Dr. George Lenchner.

SPECIFICATIONS: *Paperback. 8.5 in by 11 in. 284 pages. 2005 edition. ISBN 1-882144-10-4. Library of Congress Control Number 2005927296. Current prices are available at our Web site, www.moems.org. Add charges for shipping and handling.*

Based upon Dr. Lenchner's popular inservice course, this textbook is organized into four parts: techniques for teaching problem solving; strategies used to solve problems; problem solving within many topics in the standard curriculum; and, in the six appendices, extensions of some topics and explorations of others. The book extends elementary and middle school mathematics through approximately 400 non-routine problems into such topics as sequences, series, principles of divisibility, geometric configurations, and logic.

Creative Problem Solving in School Mathematics, 2nd Edition is a valuable resource for teachers in our program and useful to parents who want to help their children grow mathematically. Each point made in the book is supported by a wealth of challenging problems, with detailed solutions. Further, the material in the second and third parts is organized into 50 sections, each of which can be presented as a full lesson.

ORDER FORMS ARE ON PAGE **307** AND AT OUR WEB SITE, *www.moems.org.*
Purchase the full three-book problem-solving set at a discount of 10%.

MATH OLYMPIAD CONTEST PROBLEMS
VOLUME 2
Edited by Richard Kalman

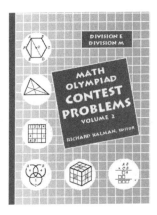

SPECIFICATIONS: *Paperback. 8.5 in by 11 in. 309 pages. 2007 edition. ISBN 1-882144-11-2. Library of Congress Catalog Number: 2007908880. Current prices are available at our Web site, www.moems.org. Add charges for shipping and handling.*

This is the book in your hands now. It is the companion book to *Math Olympiad Contest Problems for Elementary and Middle Schools* by Dr. George Lenchner and contains the 425 problems that appeared on the 85 Olympiads between 1995 and 2005. Features include detailed solutions, hints, percents correct, a problem index, and a rich study guide for mathletes and coaches.

ORDER FORMS ARE ON PAGE 307 AND AT OUR WEB SITE, *www.moems.org*. *Purchase the full three-book problem-solving set at a discount of 10%.*

MATH-TAC-TOE
AN EXCITING THINKING GAME

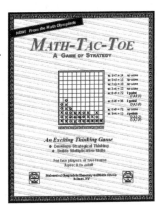

SPECIFICATIONS: *5.5 in by 8.5 in. 5 pads per set, each with 20 game boards and score cards, instructions and scoring key. Current prices are available at our Web site, www.moems.org. Prices include discounts for orders of more than one set. Charges for shipping and handling are included.*

A mathematical extension of Tic-Tac-Toe, **Math-Tac-Toe** is a fun way to develop strategical thinking while building and reinforcing multiplication skills. Students choose and place multiplication facts on the game board in order to occupy 4 boxes in a row, in as many directions as possible. Players focus on thinking offensively and defensively at the same time; multiplication becomes a necessary tool, rather than an end in itself.

Shipping and handling are included.

ORDER FORMS FOR MATH-TAC-TOE ARE ON PAGE 307 AND AT OUR WEB SITE, *www.moems.org*.

STUDENT SPIRIT ITEMS

To help their students really get into the spirit of the contests, some PICOs ask their mathletes to wear team shirts on contest days. This also helps to make the math team or math club appear more prominent in the eyes of both the staff and the student body. Other PICOs order shirts, caps, and/or pennants as gifts for team members after the last contest.

THE OLYMPIAD TEE SHIRT

Our tee shirt is royal blue with a white Olympiad Logo on the front. It is available in assorted adult sizes (small, medium, large, extra large). Because shipping and handling is included, the minimum order is 4 shirts.

Current prices are available at our Web site, www.moems.org.

THE OLYMPIAD CAP

Our cap is royal blue ▲with a white Olympiad Logo on the front. It comes in one size with an adjustable headband to fit all adults and children. Because shipping and handling is included, the minimum order is 6 caps.

Current prices are available at our Web site, www.moems.org.

THE OLYMPIAD PENNANT

Our pennant is royal blue felt with a white Olympiad Logo and white lettering. The size is 5" by 12". Because shipping and handling is included, the minimum order is 6 pennants.

Current prices are available at our Web site, www.moems.org.

ORDER FORMS FOR THESE SPIRIT ITEMS ARE ON PAGE 307 AND AT OUR WEB SITE, *www.moems.org.*

FOR TEACHERS, STUDENTS, PARENTS
AND OTHER INTERESTED PEOPLE

Ordering

Please use the current prices, available at our Web site, *www.moems.org*, for both items and shipping. Orders from USA ZIP codes must prepay using a major credit card, check, or school purchase order. All orders from other than USA postal codes must prepay with a major credit card.

Domestic prices apply to USA, Canada and Mexico postal codes. Foreign prices apply to all others. Shipping and handling fees are charged only for books. Most shipping is done through UPS Ground or Global Priority Mail.

You may use the order forms at our Web site or the combined form below.

Name _____ School _____

School Address _____

City, State, Zip _____ Phone (Day) _____

Email _____

PAYMENT: Check # _____   P.O.# _____   *(payable to **MOEMS**)*

Credit card: ❏ **MasterCard**   ❏ **Visa**   ❏ **Discover**   ❏**AMEX**

#_____ _____ _____ _____   Expires _____

Books	Quantity	Unit Price	Amount
Contest Problems for Elementary & Middle Schools			
Creative Problem Solving in School Mathematics			
Contest Problems, Volume 2			
Special price if all three books are ordered together			
		Shipping and Handling	
		Subtotal	

Math-Tac-Toe	Quantity	Unit Price	Amount
I set (5 pads)			
2-4 sets			
5 or more sets			
		Subtotal	

Other items	Quantity	Unit Price	Amount
Tee shirts ___ S ___M ___ L ___ XL			
Caps			
Pennants			
		Subtotal	

	TOTAL	

FOR SCHOOL CLUBS
AND INDIVIDUAL CLASSES

THE MATH OLYMPIAD CONTESTS

Worldwide, over 5000 schools participate each year in the Olympiads. Designed for the Valley Stream, NY schools in 1979 by the district Math Director, Dr. George Lenchner, the Olympiads grew rapidly by word of mouth. Several features contribute to its growth year after year:

1. *Interesting, meaningful problems entice students to WANT to do MORE math.*
2. *5 contests per year sustain learning and interest, and build mathematical thinking.*
3. *Teams of up to 35 students encourage as many students as possible to grow mathematically.*
4. *A school enrolling more than one team underlines the importance of math within its program.*
5. *No traveling is required.*
6. *50 practice problems help students know what to expect before the first Olympiad.*
7. *Certificates for all students and other awards for half of all students provide closure.*

Teams enroll one of two ways. Most use our password-protected Web site to report and correct all student data and scores, and to receive all contests and practice materials, while the rest use the mails. Each team enrolls in either Division E (up to grade 6) or Division M (up to grade 8). Only fulltime registered students from the same member school or homeschool association are eligible to participate on a School Team. All other teams, called District Teams or Institute Teams, are ineligible for team awards.

Mail the payment or purchase order, payable to MOEMS, with the enrollment form to MOEMS, 2154 Bellmore Avenue, Bellmore, NY 11710-5645 by September 30. Refunds are granted only to schools that withdraw by September 30. Current prices are available at our Web site, *www.moems.org*.

Enrollment Form

Please type or print.

Date _____

School _____ Team Letter (*if more than one team*) _____

School Address _____ District _____

City _____ State _____ Zip + 4 _____

School phone (___) _____ Fax (___) _____

Person in Charge of Olympiad (PICO) _____

IMPORTANT — *Check one box in each Category:*

1. I prefer to: (*a*) ❑ receive all materials and register students / report scores **ONLINE**.
   Required for password: Email Address _____

   (*b*) ❑ receive all materials and register students / report score by **SNAIL MAIL**.

2. Division: ❑ **DIVISION E** (*Gr. 4-6*) ❑ **DIVISION M** (*Gr. 7-8*)

3. Payment: ❑ Check #_____ ❑ Purchase Order #_____
   Credit Card: ❑ *VISA* ❑ *MASTERCARD* ❑ *DISCOVER* ❑ *AMERICAN EXPRESS*
   # _____ _____ _____ _____ Expires _____

Membership Fee Per Team:	Online	Snail Mail
USA, Canadian, or Mexican Zip Code.	❑ $___ US	❑ $___ US
All Other Foreign.	❑ $___ US	❑ $___ US

FOR TEACHER ORGANIZATIONS
AND MID-SIZED SCHOOL DISTRICTS

YOUR ORGANIZATION'S MATH TOURNAMENT

The main features of the Regional Math Tournament are:

1. The organization or district invites several schools to send 5-member teams to compete face-to-face at one site. Gathering 100-200 students together creates excitement and underscores the importance of mathematics.

2. The full-day tournament consists of 3 rounds (individual, team, playoff) and an awards ceremony. In the individual round, each student works alone on 10 timed short-answer problems. In the team round, all five team members work cooperatively on another 10 timed short-answer problems. In the playoff round, ties for first place are broken using up to 3 additional problems with time constraints. All problems are supplied by MOEMS and are similar to those in our books.

3. The schedule includes time after each event to discuss solutions, an important feature. MOEMS supplies detailed solutions and, where possible, multiple methods.

4. This is your organization's tournament. Everything is packaged under the name and logo of your organization or district, which you place in large print at the top of every sheet of paper. Contests, flyers, and awards all display your name. "MOEMS" only appears in small print in the copyright statements.

5. The organization or district sets, collects, and keeps all per-team fees. It pays MOEMS a moderate flat fee for the masters for the problems, solutions, and a detailed instruction manual. The MOEMS flat fee is a fraction of your organization's income that is generated by the registration fee. The fee is available at our Web site, *www.moems.org*.

6. The benefits to your organization or district are many. The tournament has the potential to: enhance your organization's reputation for promoting excellence in mathematics; introduce it to teachers and administrators unfamiliar with it; help to bring in new members; and recruit new workers.

7. The manual walks your tournament committee through every aspect of mounting a tournament, including that of recruiting workers. It also contains model flyers for enrolling teams and a scorekeeping program.

8. The MOEMS manual also addresses the many subtle decisions that need to be made, ranging from those associated with recruiting workers and teams, to those associated with setting the tournament date and site and selecting awards, and so on.

For further information, email *office@moems.org*, call 866-781-2411, or write MOEMS, 2154 Bellmore Avenue, Bellmore, NY 11710-5645. Please supply the name of your organization or district and all appropriate contact information.